T0261384

PRINCETON AERONAUTICAL
PAPERBACKS

PRINCETON UNIVERSITY PRESS · PRINCETON, N. J.

NUMBER 1

PRINCETON AERONAUTICAL

PAPERBACKS

COLEMAN duP. DONALDSON, GENERAL EDITOR

LIQUID PROPELLANT ROCKETS

BY DAVID ALTMAN

JAMES M. CARTER, S. S. PENNER, AND

MARTIN SUMMERFIELD

PRINCETON, NEW JERSEY

PRINCETON UNIVERSITY PRESS

1960

© COPYRIGHT, 1956, 1959, 1960, BY PRINCETON UNIVERSITY PRESS

L. C. CARD 60-12048

Reproduction, translation, publication, use, and dis-
posal by and for the United States Government and its
officers, agents, and employees acting within the scope
of their official duties, for Government use only, is permitted. At the expiration of ten years from the date of
publication, all rights in material contained herein first
produced under contract Nonr-03201 shall be in the
public domain.

PRINTED IN THE UNITED STATES OF AMERICA

HIGH SPEED AERODYNAMICS

AND JET PROPULSION

———————◆●◆———————

BOARD OF EDITORS

THEODORE VON KÁRMÁN, *Chairman*
HUGH L. DRYDEN
HUGH S. TAYLOR

COLEMAN DUP. DONALDSON, General Editor, 1956–
Associate Editor, 1955–1956

JOSEPH V. CHARYK, General Editor, 1952–
Associate Editor, 1949–1952

MARTIN SUMMERFIELD, General Editor, 1949–1952

RICHARD S. SNEDEKER, Associate Editor, 1955–

PRINCETON, NEW JERSEY
PRINCETON UNIVERSITY PRESS

PREFACE

The favorable response of many engineers and scientists throughout the world to those volumes of the Princeton Series on High Speed Aerodynamics and Jet Propulsion that have already been published has been most gratifying to those of us who have labored to accomplish its completion. As must happen in gathering together a large number of separate contributions from many authors, the general editor's task is brightened occasionally by the receipt of a particularly outstanding manuscript. The receipt of such a manuscript for inclusion in the Princeton Series was always an event which, while extremely gratifying to the editors in one respect, was nevertheless, in certain particular cases, a cause of some concern. In the case of some outstanding manuscripts, namely those which seemed to form a complete and self-sufficient entity within themselves, it seemed a shame to restrict their distribution by their inclusion in one of the large and hence expensive volumes of the Princeton Series.

In the last year or so, both Princeton University Press, as publishers of the Princeton Series, and I, as General Editor, have received many enquiries from persons engaged in research and from professors at some of our leading universities concerning the possibility of making available at paperback prices certain portions of the original series. Among those who actively campaigned for a wider distribution of certain portions of the Princeton Series, special mention should be made of Professor Irving Glassman of Princeton University, who made a number of helpful suggestions concerning those portions of the Series which might be of use to students were the material available at a lower price.

In answer to this demand for a wider distribution of certain portions of the Princeton Series, and because it was felt desirable to introduce the Series to a wider audience, the present Princeton Aeronautical Paperbacks series has been launched. This series will make available in small paperbacked volumes those portions of the larger Princeton Series which it is felt will be most useful to both students and research engineers. It should be pointed out that these paperbacks constitute but a very small part of the original series, the first seven published volumes of which have averaged more than 750 pages per volume.

For the sake of economy, these small books have been prepared by direct reproduction of the text from the original Princeton Series, and no attempt has been made to provide introductory material or to eliminate cross references to other portions of the original volumes. It is hoped that these editorial omissions will be more than offset by the utility and quality of the individual contributions themselves.

Coleman duP. Donaldson, General Editor

PUBLISHER'S NOTE: Other articles from later volumes of the clothbound series, *High Speed Aerodynamics and Jet Propulsion*, may be issued in similar paperback form upon completion of the original series in 1961.

CONTENTS

CONTENTS

Martin Summerfield, Department of Aeronautical Engineering, Princeton University, Princeton, New Jersey

SECTION A

HIGH TEMPERATURE EQUILIBRIUM

JAMES M. CARTER

DAVID ALTMAN

A,1. Problems in Combustion. There are two principal problems in determining the nature of high temperature combustion. The first is the determination of the nature and amount of the combustion products which may exist and also the thermodynamic properties of the mixture at one or more sets of combustion conditions.

These combustion conditions are always idealized for purposes of calculation. Thus in a rocket or jet motor, steady state conditions are assumed; combustion during starting or stopping is considered separately by approximate methods from a knowledge of steady state conditions.

In order for thermochemistry and thermodynamics to apply to combustion, either certain other simplified conditions must be assumed, or much more detailed information is required than is usually available. Thus, in effect, calculations on combustion in rocket and jet motors are made on the assumption that the combustion process occurs in a small mass element which is injected into a motor near the center of an infinite stream of identical elements and which then travels through the motor with negligible velocity and at constant pressure. The combustion chamber of the motor is assumed long enough to permit equilibrium to be attained, and the walls are assumed far enough removed to exert no effect since the element in question is surrounded by identical elements.

Similarly in explosions such as may occur in guns, or in intermittent devices, calculations are actually made for a mass element of given density, surrounded by identical elements of the same density, all in a container large enough to have no effect on the element being considered, except to keep its density constant.

It is obvious that in most cases these conditions are far from fulfilled, and that effects due to combustion speed, mixing, loss of heat to walls, and other phenomena must be considered. However, the simplified conditions assumed make calculation much simpler and do provide an upper limit for the efficiency of combustion processes. When these conditions are met, the reaction equilibria among the substances present, as discussed in I,A and C, do determine the combustion products and their thermodynamic properties.

Further simplification is obtained by assuming that combustion products are perfect gases, which is nearly true for most processes treated in rockets or jet motors. Deviations which may occur at high pressures, or in ultrafast processes, are considered in I,A and C and in other texts on thermodynamics [1,2,3].

The second problem is the determination of the heat which may be released by the combustion, the maximum combustion temperature which can be reached, or the work which may be extracted, subject to the possible compositions as determined above and to the conditions under which the combustion occurs. The usual conditions assumed are those for adiabatic combustion at constant pressure or at constant volume. Other conditions (those involving heat transfer or external work) are considerably more complicated. Discussion of these conditions has been given by Lewis and von Elbe [4] and by Hirschfelder [5].

The combustion process may be represented by

combustibles (p_1, V_1, T_1) = combustion products (p_2, V_2, T_2)

$$+ Q \text{ kilocalories} \quad (1\text{-}1)$$

with Q determined by the expression

$$E_{V_2, T_2} - E_{V_1, T_1} = -Q - \int_{p_1, V_1}^{p_2, V_2} p \, dV$$

In the above, the composition of the combustibles and the initial state are known; the composition of the products and their state, together with Q, are to be determined.

This problem may conveniently be divided into two parts: the determination of equilibrium composition and thermodynamic properties; and the determination of flame temperatures and heat released or work done for various final states of the combustion products.

A,2. Determination of Equilibrium Composition and Thermodynamic Properties. The composition and thermodynamic properties of the equilibrium products of combustion are uniquely determined by the atomic composition, the temperature, and the pressure (or volume of the system), as is shown in I,A and C. In particular, at a specified temperature and pressure they do not depend on the heats of formation of the combusting materials or on their heat of reaction. (However, these quantities will determine the temperature range over which the composition and thermodynamic properties are of interest, since they determine the initial energy of the system.)

For example, the equilibrium composition and thermodynamic properties of the combustion products of the combustible mixtures represented by $6C + 3H_2 + 4O_2$, $3C_2H_2 + 4O_2$, and $C_6H_6 + 4O_2$ are identical at any chosen temperature and pressure. Flame temperatures, heat released, or work available, differ widely because of different initial energies of the three systems. These energies are determined by the basic conservation-of-energy equations.

EQUILIBRIUM COMPOSITION. Equilibrium conditions in the combustion products can be obtained from thermodynamic data and from the total amounts of each atomic species involved. For many reactions, values of the equilibrium constants are available either from experimental measurements or as calculated values from spectroscopic data. Other equilibrium constants may be calculated from free energy data. The theory of equilibrium conditions is treated in I,A and sources of data are also included in I,A.

There is always an element of arbitrariness involved in choosing the equilibria regarded as significant. For instance, a choice frequently made is to neglect all equilibria involving species present in less than 0.01 or 0.1 per cent of the total. But this is not all-conclusive; obviously an equilibrium involving 0.01 per cent of the total moles present for a reaction with an energy change of 150 kilocalories per mole is more significant, in determining the state of the system, than an equilibrium involving 0.1 per cent of the moles for a reaction with an energy change of 10 kilocalories per mole.

A rough calculation which involves the heat given off in a combustion process resulting in products stable at ambient temperature, and the specific heats of these products, serves to set an upper limit on the combustion temperature. At this temperature, the equilibria which may exist among products involving all the atomic species present are examined. For example, the combustion of gasoline (taken as octane C_8H_{18}) with oxygen may be used. A typical reaction for rocket motors may be approximated by

$$C_8H_{18} + 6O_2 \rightarrow 8CO + 4H_2O + 5H_2 \tag{2-1}$$

(In combustion reactions involving carbon, hydrogen, and oxygen, a useful rule in estimating reaction products is, (1) to oxidize carbon to CO, (2) to use remaining oxygen to oxidize hydrogen to water, and (3) if any oxygen remains, to oxidize CO to CO_2.)

If this reaction occurred at 300°K, with the products remaining at 300°K, the heat released would be about 425 kilocalories. Cooling the products to 0°K would release an additional 35 kilocalories. Thus the total heat released can be expressed as

$$C_8H_{18} \ (l) + 6O_2 \ (g, 300°K) = 8CO \ (g) + 4H_2O \ (g) + 5H_2 \ (g, 0°K)$$
$$+ \ 460 \ kcal$$

Enthalpy tables for the products establish a temperature near 3000°K, if the entire heat release is employed in raising the temperature.

Examination of the equilibrium constants for the various reactions which might be involved shows that the water-gas reaction is such that CO_2 is present, that steam will dissociate to OH, H_2, and O_2; H_2 and O_2 will dissociate to H and O respectively, but that because of the large excess of H_2, O_2 and especially O will be present in very small amounts. Therefore an initial estimate of the species present would include CO, CO_2, H_2, H_2O, OH, H.

In the calculation of equilibrium compositions, the starting point is a series of equations; these equations will equal in number all of the unknowns to be determined.

These equations are of two kinds: the material balance equations,

$$n = \sum n_i \qquad (2\text{-}2)$$

$$(M) = \sum (n_{iM}) \cdot n_i$$

$$(N) = \sum (n_{iN}) \cdot n_i \qquad (2\text{-}3)$$

etc.,

and the equilibrium equations,

$$K_i f = \frac{f_C^c \cdot f_D^d \cdots}{f_A^a \cdot f_B^b \cdots} \qquad (2\text{-}4)$$

etc.

The first equation states that the total number of moles is equal to the sum of the number of moles of the molecular components. The second set of equations expresses the fact that the total number of gram atoms of each atomic constituent, denoted by (M), (N), etc., is distributed among the molecular components n_i, with n_{iM} atoms of M in the component i. The third set of equations are the usual equilibrium expressions for the gas reactions

$$aA + bB + \cdots \rightleftarrows cC + dD + \cdots$$

In the equilibrium equations, f is the fugacity. Conversion to the usual forms for pressure, mole fraction, etc. is made by use of appropriate multiplying factors Q_p, Q_z, etc. involving fugacity as a function of pressure or density. These relations are given in I,A and C. The following equilibrium expressions are those most frequently involved in combustion involving carbonaceous fuels. The numbering used follows that given by Hirschfelder [5], and all subsequent references to equilibrium constants or reactions follow this numbering. It should be emphasized that while the expressions are given in terms of pressure, this is accurate only insofar as the perfect gas laws apply. When appreciable deviations occur, pressures must be replaced by fugacities.

$$K_1 = \frac{p_{CO} \cdot p_{H_2O}}{p_{CO_2} \cdot p_{H_2}} \qquad K_8 = \frac{p_N}{p_{N_2}^{\frac{1}{2}}}$$

$$K_5 = \frac{p_{CO}}{p_{O_2}^{\frac{1}{2}}} \qquad K_9 = \frac{p_H}{p_{H_2}^{\frac{1}{2}}}$$

$$K_6 = \frac{p_{O_2} \cdot p_{H_2}^2}{p_{H_2O}^2} \qquad K_{10} = \frac{p_{OH} \cdot p_{H_2}^{\frac{1}{2}}}{p_{H_2O}}$$

$$K_7 = \frac{p_O \cdot p_{H_2}}{p_{H_2O}}$$

In the simplest cases, when there are only one or two atomic species and two or three molecular species, these equations may be solved simply by algebraic means. In general, however, the number and complexity of the equations make this impossible.

For the reaction given in Eq. 2-1, the equilibrium equations would be set up as follows:

$$n = n_{CO} + n_{CO_2} + n_{H_2} + n_{H_2O} + n_H + n_{OH} \qquad (2\text{-}2a)$$

$$(C) = 8 = n_{CO} + n_{CO_2}$$

$$(H) = 18 = 2n_{H_2} + 2n_{H_2O} + n_H + n_{OH} \qquad (2\text{-}3a)$$

$$(O) = 12 = n_{CO} + 2n_{CO_2} + n_{H_2O} + n_{OH}$$

and if the perfect gas laws hold, so that

$$pV = nRT, \qquad \frac{f}{p} = 1, \quad \text{and} \quad p_i = p\,\frac{n_i}{n}$$

$$K_{1p} = \frac{n_{CO} \cdot n_{H_2O}}{n_{CO_2} \cdot n_{H_2}}$$

$$K_{9p} = \left(\frac{p}{n}\right)^{\frac{1}{2}} \frac{n_H}{n_{H_2}^{\frac{1}{2}}} \qquad (2\text{-}4a)$$

$$K_{10p} = \left(\frac{p}{n}\right)^{\frac{1}{2}} \frac{n_{OH} \cdot n_{H_2}^{\frac{1}{2}}}{n_{H_2O}}$$

The above set of seven equations involving seven unknowns does not lend itself to easy algebraic solution. In this case, which is relatively simple, the algebraic solution can be obtained, but ordinarily, with more molecular species and more equations involved, such a method is virtually impossible.

Recourse must then be had to one of a number of approximate methods. These may be classified as (1) trial and error methods, (2) iterative methods, (3) graphical methods and use of published tables, and (4) punched-card or machine methods.

1. **Trial and error methods.** The straightforward method of solving for an equilibrium composition may be accomplished by the insertion into Eq. 2-4 of trial values consistent with the material balance Eq. 2-3 until all equations are satisfied. Simple rules, such as the one given previously in the example employing octane and oxygen, may be employed to assign initial trial values. This direct method is the least efficacious and is recommended only in those cases where the number of components is small and a reasonably good guess can be made of the composition.

The trial and error method, however, can effectively be applied to more complicated systems if the number of working equations are first reduced by algebraic substitutions. A general method for handling C, H, O, and N systems of up to ten components is demonstrated in the following treatment.

Let the number of moles of the components be represented by the following symbols:

$$a = n_{H_2} \qquad f = n_{N_2}$$
$$b = n_{H_2O} \qquad g = n_{NO}$$
$$c = n_{CO} \qquad h = n_{OH}$$
$$d = n_{CO_2} \qquad i = n_{H}$$
$$e = n_{O_2} \qquad j = n_{O}$$

These ten unknowns are related by means of Eq. 2-3 and 2-4 as follows:

$$\frac{bc}{ad} = K_1 \tag{2-5}$$

$$\frac{ga}{f^{\frac{1}{2}}b} = K_3 \left(\frac{n}{p}\right)^{\frac{1}{2}} \tag{2-6}$$

$$\frac{a^2 e}{b^2} = K_6 \left(\frac{n}{p}\right) \tag{2-7}$$

$$\frac{aj}{b} = K_7 \left(\frac{n}{p}\right) \tag{2-8}$$

$$\frac{i}{a^{\frac{1}{2}}} = K_9 \left(\frac{n}{p}\right)^{\frac{1}{2}} \tag{2-9}$$

$$\frac{ha^{\frac{1}{2}}}{b} = K_{10} \left(\frac{n}{p}\right)^{\frac{1}{2}} \tag{2-10}$$

$$c + d = (C) \tag{2-11}$$

$$b + c + 2d + 2e + g + h + j = (O) \tag{2-12}$$

$$2a + 2b + h + i = (H) \tag{2-13}$$

$$2f + g = (N) \tag{2-14}$$

where (C), (O), (H), and (N) represent the total number of gram atoms of the elements, n is the total number of moles in the equilibrium mixture, and p is the total pressure. The subscripts on the K's denote the specific equilibria as given in Eq. 2-4. Although n is accurately given by the equation

$$n = a + b + c + d + e + f + g + h + i + j$$

the concentrations of the components are not sensitive to small variations in n. This fact permits the choice of an approximate n value in Eq. 2-6 to 2-10 with only very little loss in accuracy.

The above ten equations may now be simplified to yield the following three working equations:

$$b = \frac{(H) - 2a - K_9(n/p)^{\frac{1}{2}}a^{\frac{1}{2}}}{2 + \frac{K_{10}(n/p)^{\frac{1}{2}}}{a^{\frac{1}{2}}}} \tag{2-15}$$

$$g = (O) - (C)\frac{K_1a + 2b}{K_1a + b} - b\left[\frac{K_{10}(n/p)^{\frac{1}{2}}}{a^{\frac{1}{2}}} + \frac{K_7(n/p)^{\frac{1}{2}}}{a}\right] - 2K_6\left(\frac{n}{p}\right)\frac{b^2}{a^2} \tag{2-16}$$

$$\frac{2g^2a^2}{K_3^2(n/p)b^2} + g - (N) = 0 \tag{2-17}$$

Equations 2-15, 2-16, and 2-17 are the three basic equations upon which the trial and error method is applied. A likely value of $a = n_{H_2}$ is chosen. Values of b and g are then obtained directly from Eq. 2-15 and 2-16. The values of a, b, and g so obtained are now substituted into Eq. 2-17. Non-zero solutions of Eq. 2-17 require new trial values of a until that equation is satisfied. Once Eq. 2-17 is satisfied, all other components are obtained simply from the individual Eq. 2-5 to 2-14.

The introduction of the component n_N, which occurs at very high temperatures, or components containing other elements such as sulfur and fluorine can easily be made by this scheme.

2. Iterative methods. Briefly, this method consists in determining (by inspection or otherwise) which species are present in largest amounts, and of solving the reduced number of equations involving these species only. The minor species are then determined from the major ones by use of the equilibrium equations, and the concentration of the major species is corrected for the presence of minor ones by using the mass balance equations. New relations among the major species are then determined, using the corrected values, and a second set of values for the minor species is calculated. This process is continued until no further change is produced by repeating the process. Obviously, the method can be continued to give as accurate values as desired. Usually the process is terminated at some predetermined limit of accuracy, such as 0.1 or 0.01 mole per cent.

In the example above, the first step would be taken by assuming that

H and OH are not present. Eq. 2-2, 2-3, and the water-gas equilibrium of Eq. 2-4 are then solved algebraically for the major components. This involves only a simple quadratic, since p and n do not enter into the equilibrium equation. H and OH are then calculated from the second and third equations in Eq. 2-4. Then second approximations to (H) and (O) are calculated from Eq. 2-3 by the relation $(H)' = (H) - n_H - n_{OH}$ and $(O)' = (O) - n_{OH}$. These second approximate values are used as in the first step to calculate new values for the major components, and these in turn are used again for H and OH.

As a final step, equilibria which may have been neglected in setting up the equations are examined. (In the present case the dissociation of steam to hydrogen and oxygen would be most probable.) If the amount of oxygen does prove negligible, based on the equilibrium values found for steam and hydrogen above, the procedure is justified. However, if the amount of oxygen is comparable to the amounts of other minor constituents, Eq. 2-2 to 2-4 must be rewritten to include this component, and the process must be gone through again.

The process above is comparatively easy when the combustion mixture is relatively far from stoichiometric proportions. Fortunately this is true for most rocket propellants, which are usually underoxidized, and for fuel-air mixtures in turbojets, etc., which are overoxidized. This means that several molecular species are present in large amounts, which condition will be changed little by the presence of minor species. The iteration process therefore converges very rapidly.

As stoichiometric proportions are approached in a combustion mixture, however, the iteration process becomes progressively less efficient, particularly at the higher temperatures involved. Thus in hydrogen-oxygen mixtures, if either excess hydrogen or excess oxygen is present, it can safely be assumed that H_2O and either H_2 or O_2 are the two major components, and the concentrations of OH, H, O, and either O_2 or H_2 can be fairly easily determined.

In stoichiometric proportions the concentrations of many or all species other than H_2O are roughly equal, and a change in one introduces changes in the others which are proportionately of the same magnitude. Under these conditions, the iterative process converges very slowly. For exact stoichiometric proportions as in steam, carbon dioxide, and some other important substances, the labor involved in calculation is not too serious, since only one final composition is involved. Most combustion mixtures are not in exact stoichiometric proportions and also involve more than one combustion reaction.

Brinkley and others [6,7,8] have devised a number of systematized methods of iteration. Some of the more complex methods will even converge fairly rapidly [9, p. 187] for highly dissociated systems near stoichiometric proportions.

One method (Sachsel, et al. [*10*, p. 620]), which is applicable when electrical computers are available, is to transform all of the equilibrium equations, Eq. 2-4, to logarithmic form. These then become simple linear equations and amenable to machine calculations.

3. Graphical methods and tables for use within definite composition ranges for several important systems have been developed by Winternitz, Huff, Calvert, and Kassner [*10*, p. 620; *11*; *12*]. These, together with (4) machine and punched-card methods of computation, are more fully described in Sec. C of this volume.

From the above, it is apparent that at the combustion temperatures usually encountered, determination of equilibrium compositions is a complicated and tedious process. Even if the only information required is the composition at the flame temperature, two or three such composition calculations must be made, since an initial estimate of the flame temperature (for which the composition is required) will usually be in error by an appreciable amount. If compositions are required not only at flame temperatures over a range of pressures or densities, but at lower temperatures and pressures in order to follow expansion processes, the number of calculations which must be made is greatly increased.

Interpolation of equilibrium compositions. Frequently it is desired to know the equilibrium compositions at temperatures or pressures other than those for which calculations have been made.[1] For different temperatures, the most accurate method is the use of the equilibrium constants for the temperature in question. These can be interpolated quite accurately, since in general a plot of $\ln K$ vs. T is nearly linear. When compositions are determined at other pressures, a new set of equilibrium calculations must be made since the pressure enters into Eq. 2-4 and these equations must, therefore, be altered for the new pressure.

There is, however, a simple graphical method of obtaining equilibrium compositions at temperatures or pressures intermediate to those for which two or more calculations have been made. It can also be used for a limited amount of extrapolation. When the amounts of the various components are plotted on the logarithmic scale of semilogarithmic paper against either temperature (at one pressure) or pressure (at one temperature), it is found that nearly straight lines are obtained. Therefore the procedure is to plot the amounts of the minimum number of components (starting with the least abundant) which will completely determine the composition, on semilogarithmic paper, and to read off the amounts at intermediate temperatures or pressures. By plotting the components least abundant, the maximum accuracy is obtained, since these components undergo the greatest percentage changes with temperature or pressure. The amounts

[1] There is a definite need for sets of tables or charts for the more important systems, at reasonably close intervals of composition, and all based on the *same data* for equilibrium constants, equations of state, etc. Such a basis might be the Bureau of Standards Tables of Chemical Thermodynamic Quantities [*13*].

of the other (major) components are then determined from Eq. 2-3, and the total number of moles of gas from Eq. 2-2. Even in cases where this method does not give the accuracy desired, it can be used to eliminate all but the final steps in an exact calculation by the iterative method.

THERMODYNAMIC PROPERTIES. As in the case of the compositions, the thermodynamic properties of gases are point functions of atomic composition, temperature, and pressure or volume. However, for energy quantities there is the requirement of a base point. The choice of this point may be entirely arbitrary, but once the choice is made all thermodynamic quantities must be referred to it. Several such bases are in common use. Two bases frequently employed are the system stable at room temperature (25°C, 300°K, 70°F) and at a pressure of one atmosphere, and the system hypothetically stable at 0°K and one atmosphere, with combustion products ordinarily gaseous in the perfect gas state at this temperature. Since more and more tables of enthalpies, entropies, etc., based on 0°K are becoming available, this latter base has much to recommend it. (See Sec. C and I,A.)

For calculations involving shifts in chemical equilibrium during expansion, two other bases are more logical. The first is a base with the elements at ambient temperature, and the second, with the elements at 0°K. The latter has the advantage of corresponding more closely with theoretical considerations for absolute enthalpy, entropy, etc. It has the practical disadvantage at present that many heats of formation at 0°K are not known, and that in some cases enthalpy differences between ambient conditions and 0°K are not known. However, this situation is being remedied rapidly [13]. In the following discussion, the use of this base is assumed, as it results in the simpler formulas. If some other base is used, a constant is added to all thermodynamic quantities. Since behavior on combustion and expansion always involves differences, this constant cancels out in any computation.

The determination of enthalpy and internal energy for the equilibrium compositions at various temperatures and pressures (or densities) may be divided into two parts. The first is that due to the heat or energy of reaction at the base temperature, from the standard composition at base temperatures to the equilibrium composition at the temperature T. The second is the change in enthalpy or internal energy in heating this mixture from the base temperature to the temperature T, plus the effects in going from the perfect gas state at one atmosphere to the pressure or density under consideration. These relations are expressed in the equations below.

$$H = \sum n_i \cdot (\Delta H_f^0)_i + \sum n_i (H_T^0 - H_0^0)_i + n \int_{p_0}^{p} \left(\frac{\partial H}{\partial p}\right)_T dp \quad (2\text{-}18)$$

$$E = \sum n_i \cdot (\Delta E_f^0)_i + \sum n_i (E_T^0 - E_0^0)_i + n \int_{V_0}^{V} \left(\frac{\partial E}{\partial V}\right)_T dV \quad (2\text{-}19)$$

Here, n_i is the number of moles of component i present at temperature T and ΔH_f^0 or ΔE_f^0 is the heat or energy of formation of the component i at 0°K, from the elements at 0°K. $H_T^0 - H_0^0$ or $E_T^0 - E_0^0$ is the integral of the specific heat of component i from 0°K to T.

If the elements at ambient temperature (298°K) are taken as a base point, Eq. 2-18 becomes

$$H = \sum n_i \cdot [\Delta H_{f(298)}^0]_i + \sum n(H_T^0 - H_{298}^0)_i + \cdots \qquad (2\text{-}18a)$$

and similarly for Eq. 2-19. If a base of stable composition at ambient temperature is used, the first term on the right of Eq. 2-18a is replaced by $\sum \Delta n_i \Delta H_{f(298)}^0$, where Δn_i is the change in component i in going from the composition stable at 298°K to the equilibrium composition at T. (It should be noted as an advantage of the base at 0°K that $\Delta H_f^0 = \Delta E_f^0$ and $H_0^0 = E_0^0$.) The last terms in the equations correct for the changes in enthalpy or internal energy with pressure or volume. For perfect gases these terms are zero. In the general case, as listed by Bridgman [14] for these and other thermodynamic formulas,

$$\left(\frac{\partial H}{\partial p}\right)_T = V - T\left(\frac{\partial V}{\partial T}\right)_p \qquad (2\text{-}20)$$

$$\left(\frac{\partial E}{\partial V}\right)_T = T\left(\frac{\partial p}{\partial T}\right)_V - p \qquad (2\text{-}21)$$

The numerical values of the expressions above are determined from the equation of state for the gas mixture.

Entropy values for combustible mixtures may be determined in a number of ways. For many substances entropy tables are available for the standard state at a pressure of one atmosphere [13]. The entropy of the mixture is then given by the expression

$$S^0 = \sum n_i S_i^0 - nR \sum X_i \ln X_i, \qquad X_i = \frac{n_i}{n} \qquad (2\text{-}22)$$

For other materials, tables of the free energy function are available [13]. By use of the relation $F = H - TS$, the entropy may be determined as

$$S^0 = \frac{1}{T} \sum n_i (H_T^0 \cdot H_0^0)_i - \sum n_i \left(\frac{F^0 - H_0^0}{T}\right)_i - nR \sum X_i \ln X_i \qquad (2\text{-}23)$$

In both the above equations, the term $nR\sum X_i \ln X_i$ is the entropy of mixing all the components, each originally present at one atmosphere, to give the mixture at one atmosphere.

To obtain the entropy at the desired pressure or volume, there must be added to the above values the change of entropy with pressure or volume. This is given by the equation

$$S - S^0 = - \int_{p_0}^{p} \left(\frac{\partial V}{\partial T}\right)_p dp = \int_{V_0}^{V} \left(\frac{\partial p}{\partial T}\right)_V dV \qquad (2\text{-}24)$$

For most of the applications met with in jet and rocket propulsion, gases may be assumed to obey the perfect gas laws, so that the above equations reduce to

$$S - S^0 = -nR \ln \frac{p}{p_0} = nR \ln \frac{V}{V_0} \qquad (2\text{-}25)$$

Specific heats for the combustion products are obtained from the expressions

$$C_p = \left(\frac{\partial H}{\partial T}\right)_p, \qquad C_V = \left(\frac{\partial E}{\partial T}\right)_V \qquad (2\text{-}26)$$

In the above, it is necessary to keep in mind that the changes in enthalpy or internal energy include not only the amounts due to change in temperature for each species, but also those due to the change in equilibrium composition with temperature, according to the equations

$$\left(\frac{\partial H}{\partial T}\right)_p = \sum \left(\frac{\partial n_i}{\partial T}\right)_p (\Delta H_f^0)_i + \sum \left(\frac{\partial n_i}{\partial T}\right)_p (H_T^0 - H_0^0)_i$$

$$+ \sum n_i \left(\frac{\partial (H_T^0 - H_0^0)_i}{\partial T}\right)_p - T \int_{p_0}^{p} \left(\frac{\partial^2 V}{\partial T^2}\right)_p dp \quad (2\text{-}27)$$

$$\left(\frac{\partial E}{\partial T}\right)_V = \sum \left(\frac{\partial n_i}{\partial T}\right)_V (\Delta E_f^0)_i + \sum \left(\frac{\partial n_i}{\partial T}\right)_V (E_T^0 - E_0^0)_i$$

$$+ \sum n_i \left(\frac{\partial (E_T^0 - E_0^0)_i}{\partial T}\right)_V + T \int_{V_0}^{V} \left(\frac{\partial^2 p}{\partial T^2}\right)_V dV \quad (2\text{-}28)$$

The first two terms in the above give the heat involved in changes in composition, the third terms give the perfect gas value of the specific heat at fixed composition, and the last terms give the effect due to gas imperfections.

In view of the complex nature of these equations, a convenient method of obtaining average values for the specific heats over a range of temperatures is by differences from tables of enthalpy or internal energy, if these values are determined at fairly close intervals, both in temperature and in pressure (or volume).

For some purposes, $\gamma = C_p/C_V$ is required. This can be obtained from enthalpy and energy tables as above. For more accurate determinations, use is made of the thermodynamic relationship

$$C_p - C_V = -T \left(\frac{\partial V}{\partial T}\right)_p^2 \left(\frac{\partial p}{\partial V}\right)_T \qquad (2\text{-}29)$$

This can be cast into the form

$$\frac{1}{\gamma} = \frac{C_V}{C_p} = 1 - \frac{T}{C_p}\left(\frac{\partial p}{\partial T}\right)_V\left(\frac{\partial V}{\partial T}\right)_p \qquad (2\text{-}30)$$

where C_p is the heat capacity at constant pressure, T is the absolute temperature, V is the volume, and p is the pressure. All quantities must be expressed in consistent units. The partial derivatives are obtained from the compositions and the equation of state for the particular mixture under consideration. In particular for nonreacting mixtures obeying the perfect gas law the above expression reduces to

$$\frac{C_V}{C_p} = 1 - \frac{R}{C_p} \qquad (2\text{-}31)$$

Another thermodynamic property which is useful in dealing with propellants is the velocity of sound. This is given by the expression

$$a^2 = \left(\frac{\partial p}{\partial \rho}\right)_s \qquad (2\text{-}32)$$

It is shown in thermodynamics texts [1] that this reduces to

$$a^2 = \frac{C_p}{C_V}\left(\frac{\partial p}{\partial \rho}\right)_T \qquad (2\text{-}33)$$

Here again all of the required quantities may be obtained from those previously calculated and from the equation of state of the gas mixture.

The extent of the calculations made for composition, equation of state, and thermodynamic properties will vary with the information desired. This may range from a fairly rough estimate of performance characteristics for one particular mixture at one pressure or density, to detailed calculations for a range of mixtures, at a number of pressures or densities, expanded over a series of expansion conditions. The latter in effect requires the preparation of a number of Mollier or equivalent charts, one for each mixture investigated. If anything more than a single performance value is required, it is surprising to one who has not worked with the calculations before to see how little additional effort is required to cover a wide range of combustion and expansion conditions.

A,3. Determination of Heat Release and Flame Temperature. As has been stressed above, the equilibrium molecular composition and thermodynamic properties of any particular atomic composition do not depend on the initial state of the mixture. In other words, the particular nature, heats of formation, etc., of a combustible mixture have no effect on the ultimate composition or thermodynamic properties of the mixture after combustion, when these properties are taken as functions of the temperature and pressure or density.

However, it is just these factors which determine the heat released in a combustion and the flame temperature which may be reached. To evaluate these effects it is necessary to return to Eq. 1-1 and to separate this into two parts. Eq. 1-1 reads:

combustibles (p_1, T_1, V_1) = combustion products (p_2, T_2, V_2)

$$+ Q \text{ kilocalories}$$

This may be separated into two steps:

combustibles $(p_1, T_1, V_1) \rightarrow$ products, standard (p_0, T_0, V_0)

$$+ Q_0 \text{ kilocalories} \quad (3\text{-}1)$$

and

products, standard $(p_0, T_0, V_0) \rightarrow$ products (p_2, T_2, V_2)

$$+ Q_2 \text{ kilocalories} \quad (3\text{-}2)$$

In the preceding articles, the thermodynamic quantities involved in the second of these processes have been discussed. Q_2 is simply the negative of the enthalpy or internal energy at p_2, T_2, V_2. All that is required to determine performance, flame temperatures, and other required data is to evaluate the thermodynamic quantities involved in the first process.

The heat or energy values involved in the first process above will of course depend on the thermochemical properties of the combustibles, on their initial state, and on the base taken for p_0, T_0, V_0. It is here that the advantage of the base taken at ambient conditions is most apparent. In such a case Q_0 is merely the heat or internal energy involved in the combustion at constant temperature. If the base at 0°K is taken it is necessary to add to the usual heat or energy of combustion the heat involved in cooling the combustion products from ambient temperature to 0°K.

Several specific processes may be mentioned in relation to the above. Maximum flame temperature is determined by the relation $Q_0 + Q_2 = 0$. At either constant pressure or volume, if all of the heat or energy released in going from the combustibles to the base state is absorbed in going from the base state to the combustion conditions, no heat or energy is lost in the process, and the maximum temperature is attained. Several examples are given in Sec. L and M.

Furthermore, the heat or energy converted to kinetic energy or work is directly obtainable from the above equations. If the final state of the combustion products has been determined after expansion from the combustion conditions as shown in Sec. B, the extent of energy conversion is determined by

$$W = Q_0 + Q_2$$

where Q_2 denotes the heat or energy involved in going from standard conditions to final condition. It should be noticed that this equation applies not only to expansion in which equilibrium composition of the combustion products is maintained during expansion, but also when the composition is assumed frozen, either at expansion conditions or at some intermediate temperature, since Q_2 is determined by composition as well as by temperature, according to Eq. 2-18 and 2-19.

A,4. Gas Imperfection. The effect of gas imperfection in modifying equilibrium compositions and thermodynamic properties has been indicated in some of the equations above. For actual calculations involving gas imperfection, it is necessary to employ an equation of state for the gases involved. By the use of such an equation, the relations between pressure, volume, and temperature of a gas are determined over the range of interest, and corrections can be applied to the equilibrium constants and to the thermodynamic properties.

Equations of state. There are various equations of state which have been proposed to correct for deviations from the perfect gas laws. Among these are the van der Waals equation [15], Berthelot's equation [16], the Beattie-Bridgeman equation [17], the Hirschfelder equation [5], particularly adapted to high temperature and pressure, and numerous more or less empirical equations. Graphical representations have also been employed. These are treated in I,C.

A general representation of an equation of state for a pure gas may be given by

$$pV = RT\left(1 + \frac{b}{V} + \frac{c}{V^2} + \frac{d}{V^3} + \cdots\right) \qquad (4\text{-}1)$$

where b, c, d, etc. may be functions of temperature. The values of b, c, d, etc., are determined by the nature of the gas. They are usually determined from experimental measurements, although in some cases they may be calculated from theoretical considerations.

For a single pure gas, Eq. 4-1 shows a definite relationship involving temperature, volume, and pressure. Furthermore, the fugacity, on which corrections to the equilibrium constants depend, can be calculated for a pure gas.

The exact definition of fugacity and consideration of its application is given in I,A. It is sufficient to note here that fugacity is determined by the expression

$$\ln\frac{f_2}{f_1} = \frac{1}{RT}\int_{p_1}^{p_2} V\,dp \quad (T = \text{const}) \qquad (4\text{-}2)$$

It can be seen that for a perfect gas $f_2/f_1 = p_2/p_1$, but that in general for imperfect gases this is not true.

$$\lim_{p_1 \to 0} p_1 = f_1$$

Real difficulty arises with gas mixtures, both in determining p, V, T relationships and to a greater extent in determining fugacity. Experimental data for the p, V, T behavior of gas mixtures is not abundant, and in any case the entire range of compositions cannot be covered. It is not immediately apparent how the constants in Eq. 4-1 for pure gases should be combined to give an equation for a gas mixture. Further, since fugacities are usually required for individual components of a mixture, there is difficulty in choosing the quantities and limits in the integral in Eq. 4-2.

Early attempts to compute the p, V, T relationships of gas mixtures were expressed in Dalton's and in Amagat's laws. Dalton's law states that the pressure exerted by a gas mixture is the sum of the individual pressures each gas would exert if it occupied the total volume alone

$$p = n_1 \cdot p_1(T, V) + n_2 \cdot p_2(T, V) + \cdots \qquad (4\text{-}3)$$

Amagat's law states that the volume occupied by a gas mixture is the sum of the individual volumes which would be occupied by each gas alone at the same total pressure:

$$V = n_1 \cdot V_1(T, p) + n_2 \cdot V_2(T, p) + \cdots \qquad (4\text{-}4)$$

For perfect gases, both laws are true. Amagat's law is in general somewhat more accurate, but neither is a good approximation at even moderate pressures. Closer approximations require methods for combining the constants in the equation of state Eq. 4-1 for pure gases to represent the behavior for gas mixtures. Consideration of these questions is given in I,C.

Evaluation of constants. Constants in the equations of state may be evaluated in various ways. A series of p, V, T measurements over a wide range furnishes data for evaluation of the constants; they can also be evaluated from critical data and from data on intermolecular forces.

A fairly extensive treatment of the calculation of the constants of the equation of state for combustion gases is given by Hirschfelder, et al [5]. The equation of state for gases at high temperatures adopted for this purpose is a somewhat simplified van der Waals equation:

$$\frac{pV}{RT} = 1 + \frac{b}{V} + \frac{0.625b^2}{V^2} + \frac{0.2869b^3}{V^3} + \frac{0.1928b^4}{V^4} \qquad (4\text{-}5)$$

which at high densities merges into the limiting form

$$\frac{pV}{RT} = \left[1 - 0.6962 \left(\frac{b}{V}\right)^{\frac{1}{3}}\right]^{-1} \qquad (4\text{-}6)$$

In these equations attractive forces between molecules are disregarded (terms involving a in the van der Waals equation). The numerical coefficients are derived theoretically for the overlapping of rigid spheres. In the range of combustion temperatures (1500 to 5000°K) values of b are nearly independent of temperature and are assumed constant. Values of b for individual gases are evaluated from theoretical considerations on intermolecular forces. For gas mixtures, values of b may be calculated approximately at moderate pressures according to the relation

$$b = \sum X_i b_i$$

where X_i is the mole fraction of component i. Fowler [18] gives considerations for calculation of constants involving interaction between different molecular species. With such simple equations as those above, the p, V, T relations for gas mixtures are obtained rather simply. Also, the changes in enthalpy, internal energy, and entropy with pressure or density at constant temperature are readily obtained by the usual thermodynamic formulas:

$$\left(\frac{\partial H}{\partial p}\right)_T = V - T\left(\frac{\partial V}{\partial T}\right)_p \tag{2-20}$$

$$\left(\frac{\partial E}{\partial \bar{V}}\right)_T = T\left(\frac{\partial p}{\partial T}\right)_V - p \tag{2-21}$$

$$\left(\frac{\partial S}{\partial p}\right)_T = -\left(\frac{\partial V}{\partial T}\right)_p \tag{4-7}$$

$$\left(\frac{\partial S}{\partial \bar{V}}\right)_T = \left(\frac{\partial p}{\partial T}\right)_V \tag{4-8}$$

Changes in equilibrium constants with pressure or density are, however, more complicated since the fugacities of the individual components are involved. To obtain reasonably simple results, it is convenient to assume that a gas mixture is a perfect solution [19] or that

$$\frac{pV}{RT} - 1 = \sum_i \left[\frac{pV}{RT} - 1\right]_i X_i \tag{4-9}$$

By use of this assumption and the equation of state, activity coefficients and fugacities for the components in a mixture can be calculated. A detailed description of the method of calculation and tabular values of activity coefficients for the common gases are given in reference [5].

One difficulty in the application of the corrections above to the calculation of thermodynamic properties, or even to the calculation of p, V, T relationships, is that all of these equations apply only to gases of constant composition. That is, all of the partial derivations above should contain another symbol to show constant composition such as $(\partial H/\partial p)_{T,n}$ and all other quantities also should be understood to refer to a gas of fixed composition. When the deviations from the perfect gas laws are appreciable, this requires a large number of calculations. Thus in obtaining the enthalpy of an equilibrium mixture at one temperature for a series of pressures, it is not enough to integrate one equation of the form of Eq. 2-7 at a series of pressures. For each pressure a different expression must be used, since the equilibrium composition and hence the equation of state will be different for each pressure.

Effect of deviations from perfect gas behavior. It is fortunate that in most of the applications required in jet and rocket propulsion, gas mixtures may be considered to obey the perfect gas laws.

Deviations from perfect gas behavior affect both combustion conditions and thermodynamic properties. Combustion conditions are affected largely by fugacity changes, which may be appreciable when pressures are high. Thus the water-gas equilibrium is properly written

$$K_{1f} = \frac{f_{CO} \cdot f_{H_2O}}{f_{CO_2} \cdot f_{H_2}} \qquad (4\text{-}10)$$

where f is fugacity and is equal to pressure only at low pressures.

This may be altered to the usual form involving partial pressures by introducing activity coefficients a, where $a = f/p$ (see Brinkley and Kirkwood, [10, p. 586]). Then the water-gas equilibrium expression becomes

$$K_{1p} \cdot G_1 = \frac{p_{CO} \cdot p_{H_2O}}{p_{CO_2} \cdot p_{H_2}}, \qquad G_1 = \frac{a_{CO_2} \cdot a_{H_2}}{a_{CO} \cdot a_{H_2O}} \qquad (4\text{-}11)$$

If all activities are assumed unity then $G_1 = 1$, and the usual equilibrium expression is obtained. At high gas densities, deviations from the gas laws alter activity coefficients markedly, and the values of G may depart widely from unity. Values of G for a number of equilibria are given in [5] for a number of conditions. A few are given in the table below for illustration of the magnitudes involved.

Table A,4a. Corrections to equilibrium constants.

$\dfrac{V_0(273°K,\ 1\ atm)}{V(T,\ p)}$	G_1^*	$G_1^2 G_6$	G_6	G_9
50	1.090	1.090	0.916	1.010
100	1.223	1.241	0.830	1.021
200	1.736	1.991	0.661	1.045
500	17.78	74.7	0.237	1.147
1000	11,670 (3000°K)	1.3×10^6 (3000°K)	0.00956	1.476

* G_1 is somewhat affected by temperature at the highest densities.

Water-gas equilibrium $\qquad\qquad K_1 G_1 = \dfrac{p_{CO} \cdot p_{H_2O}}{p_{CO_2} \cdot p_{H_2}}$

Dissociation of carbon dioxide $\quad K_1^2 K_6 G_1^2 G_6 = \dfrac{p_{CO}^2 \cdot p_{O_2}}{p_{CO_2}^2}$

Dissociation of steam $\qquad\qquad K_6 G_6 = \dfrac{p_{H_2}^1 \cdot p_{O_2}}{p_{H_2O}^2}$

Dissociation of hydrogen $\qquad\quad K_9 G_9 = \dfrac{p_H}{p_{H_2}^{\frac{1}{2}}}$

The figures above show that increase of density has a pronounced effect in throwing the water-gas equilibrium to the side of carbon monoxide and steam. If the gases were perfect this equilibrium would remain unaffected by density or pressure. The dissociation of carbon dioxide is increased manyfold over what would occur for perfect gases, so much so that at the highest densities, dissociation *increases* with pressure.

On the other hand, dissociation of water is suppressed below what would be expected, and dissociation of hydrogen is little affected.

The thermodynamic properties are much less affected by the deviations from the gas laws, but the effects are appreciable. The principal quantities to be considered are the internal energy, the enthalpy, and the entropy. It has been found both experimentally and from the equations of state, that internal energy varies nearly linearly with the *density*, and that $\Delta(pV)$, and hence enthalpy, and the entropy difference from perfect gas behavior also vary almost linearly with density up to moderate density values. This near-linear behavior does not hold if these deviations are plotted against pressure. An indication of the magnitude of the deviations involved is given below.

Table A,4b. Deviations from gas laws; $T = 3000°K$, $V = V(1 \text{ atm})/100$.

Gas	pV/RT	p, atm	$\Delta(pV)$, cal/mole	ΔE, cal/mole	ΔH, cal/mole	$\delta\Delta S^*$, cal/mole°C
H_2	1.060	106	+35.8	+5.1	+40.9	−0.1003
N_2	1.155	116	+92.4	+3.5	+95.9	−0.2804
CO_2	1.311	131	+186.0	−18.2	+168.0	−0.6265
H_2O	1.043	104	+25.6	−20.0	+5.6	−0.1522

* $\delta\Delta S = \Delta S - R \ln (V/V_0)$

The relative effect of these deviations can be judged from the fact that C_V at this temperature is some 10 to 12 calories per mole per degree; flame temperatures are therefore altered only a few degrees by direct effects of deviations. Temperature alterations may be much greater because of composition changes induced by fugacity. Similarly, the ideal entropy change for a volume change by a ratio of 100 is about 9 entropy units; even the largest deviation is only 7 per cent of this.

A,5. Failure to Maintain Equilibrium in Combustion.

Effect of incomplete combustion. All of the above discussion has been predicated on reaching an equilibrium condition in the combustion process. This is not always so, and except in a few cases combustion is never complete. There are a number of factors which may limit the completion of combustion. In rocket motors, they may be placed into two general classes, the first involving factors limiting complete mixing, and the second involving the relation between reaction rates and the time that materials remain in the combustion chamber. Among the factors affecting mixing may be listed injector configuration, spray patterns, and the configuration of the combustion volume. Factors involving the time include L^*, and the absolute size of the combustion volume. It is apparent that if a rocket or jet motor is scaled up geometrically, the time from injection to reaching the expansion nozzle is increased. The effect of these factors is given in XI,B and H, and in XII,G and H.

In closed bombs, and to a first approximation in guns [4], the completeness of combustion may be affected by the presence of massive cold walls, by lack of turbulence, and by the extremely short time interval available for combustion.

The evaluation of the effects of incomplete combustion is quite difficult. In the first place, either extensive experimental data on the composition of the combustion products is required, or more or less ad hoc *assumptions must be made.* For these assumptions, several choices are possible: part of the fuel may be assumed to be completely unreacted and only heated, or it may be more or less decomposed without combustion; part of the oxidizing material may be similarly considered; or the combustion may be assumed to proceed to more or less arbitrary intermediate products. Consideration of possible reaction processes is given in Sec. D and E.

There is little or no difficulty involved in the determination of thermodynamic properties, heat release, or flame temperature when a composition arising from combustion has been chosen. The principal difficulty in this case is in deciding which composition most nearly approaches the actual course of combustion.

Equilibrium lag. Somewhat allied to incomplete combustion, but more amenable to theoretical treatment, is the question of lag in equilibrium. This may arise from several factors, among which may be mentioned the persistence of metastable states, failure of internal energy states to participate in the combustion process, and detonation phenomena. Examples of the first will be recalled by anyone who has seen a rocket motor operate with a nitric acid oxidizer below full efficiency. Equilibrium conditions in combustion require the presence of only minute amounts of the oxides of nitrogen, but the yellow exhaust is strong evidence for the presence of large amounts of NO or NO_2. Similarly, the odors arising from the use of organic fuels in rocket motors indicate the presence of intermediate products, since the equilibrium combustion products CO and CO_2 are odorless.

There is less evidence for the failure of internal energy states to participate in the combustion equilibrium, as shown in I,H. Some experimental evidence that rotational and vibrational states may not always be in equilibrium has been obtained by means of extremely minute orifices or venturi tubes used in obtaining gas samples. However, for most combustion processes in jet and rocket motors, the scale is many times larger than for sampling devices, so that the times involved are longer by several orders of magnitude. In addition, water vapor is present in most combustion products and has a pronounced catalytic effect in promoting vibrational and rotational equilibrium. Recent experimental work has shown that the relaxation time in pure CO_2 for vibrational energy is some 10^{-4} seconds or longer, while the presence of small amounts of water vapor reduced this to 10^{-8} seconds [20].

Detonation phenomena. In combustion processes having to do with propulsion, detonations are unwanted and indeed may lead to failure of equipment. The one exception to this is in the use of detonations for the production of high speed (ca. 20,000 ft/sec) metallic jets, as in the well-known bazooka charge.

Ordinarily in propulsion equipment, detonations occur only by accident. In jet motors and liquid fuel rocket motors, they may arise from ignition failure which permits the combustion chamber to be filled with a detonable mixture. In solid fuel rocket motors, breakup or fracture of the solid propellant will lead to rapid burning or deflagration which may under certain conditions transform into detonation.

Damage from detonations is due to two causes, (1) the extreme rapidity of combustion leading to high static pressures because of the large amount of combustion products present, and (2) the shock waves which are always present.

Combustion equilibrium as discussed above plays no part in detonation processes. The stability of the detonation is dependent upon a sharp discontinuity which travels through the mass at extreme speeds. Furthermore, in the region of the shock wave combustion products are moving at high velocities, of the order of half the velocity of the wave itself (see Brinkley and Kirkwood [10, p. 586]).

As a consequence of the great velocity of the detonation wave and the nearly as great "stream velocity" of the combustion products, there is little opportunity for any but translational degrees of freedom to be involved. Thus it can be shown (see Ubbelohde, [*10*, p. 566]) that with known detonation velocities and thicknesses of the wave region, a molecule can make *only a few collisions* (3 to 6) before the wave has passed. Since the combustion products then act as if their specific heats were those of a monatomic gas, the temperatures reached in a detonation wave are very high, even when allowance is made for the energy used in producing the "stream velocity" of the products. It is also highly probable that a large proportion of the combustion products are in excited states.

Behind the detonation wave, conditions are more calculable on an equilibrium basis, since the rotational and vibrational degrees of freedom have time to "catch up" with the rest of the process, and molecules lose their energy of excitation.

The conditions of detonation are based upon the Rankine-Hugoniot [*21*] equations for the continuity of mass, energy, and momentum across the detonation wave. These equations give the "dynamic adiabatic" of Hugoniot, but do not predict any one stable detonation velocity. This velocity is determined by application of the Chapman-Jouguet condition [*22,23*] which states that the stable velocity is that given by drawing a tangent to the Hugoniot curve in the p, V plane from the initial condition point p_0, V_0. There is recent work by Brinkley and others to indicate that the Chapman-Jouguet hypothesis concerning detonation velocity may be proved by making some reasonable assumptions (see Brinkley and Kirkwood, [*10*, p. 586]; also [*24*]). For further discussion of conditions of detonation, see III,D.

One of the principal difficulties in obtaining good numerical results from the above lies in lack of knowledge of the equation of state for combustion products. With initially gaseous materials no serious difficulties are encountered, but with solid explosives having densities equal to or greater than that of water, detonation pressures may be of the order of 10^5 atm, and the nature of the equation of state is uncertain. Unfortunately, as has been shown by Jones [*10*, p. 590], accurate measurements of detonation phenomena, used in conjunction with the Hugoniot relations, do not conversely serve to determine an equation of state. This is because temperatures are not included in the equations (only p and V). The means used to calculate temperature also influence the equation of state which is derived.

The phenomena of detonation are treated in greater detail in Sec. N, in III,D and III,G, and in the references given above, some of which contain fairly extensive bibliographies.

A,6. Cited References and Bibliography.

Cited References

1. Glasstone, S. *Textbook of Physical Chemistry*. Van Nostrand, 1946.
2. Epstein, P. S. *Textbook of Thermodynamics*. Wiley, 1937.

3. Steiner, L. *Introduction to Chemical Thermodynamics*. McGraw-Hill, 1941.
4. Lewis, B., and von Elbe, G. *Combustion, Flames and Explosion of Gases*. Macmillan, 1938.
5. Hirschfelder, J. O., et al. *Natl. Defense Research Comm. Rept. A-116*, 1942.
6. Brinkley, S. R., Jr., *J. Chem. Phys. 14*, 563 (1946); *15*, 107 (1947).
7. Krieger, F. J., and White, W. B. *J. Chem. Phys. 16*, 358 (1948).
8. Damköhler, G., and Edse, R. *Z. Elektrochem. 49*, 178 (1943).
9. Scarborough, J. B. *Numerical Mathematical Analysis*. Johns Hopkins Univ. Press, 1930.
10. *Third Symposium on Combustion, Flame, and Explosion Phenomena*. Williams & Wilkins, 1949.
11. Huff, V. N., and Calvert, C. S. *NACA Tech. Note 1653*, 1948.
12. Kassner, R. *Tech. Rept. GS-USAF (No. F-TR-2185 ND)*, 1948.
13. Rossini, F. D., Wagman, D. D., Evans, W. H., Levine, S., and Jaffe, I. Selected values of chemical thermodynamic properties. *Natl. Bur. Standards Circ. 500*, 1952.
14. Bridgman, P. W. *Phys. Rev. 3*, 273 (1914).
15. van der Waals, J. D. *Die Kontinuität des gasförmigen und flüssigen Zustandes, 1. Bd.*, 2. Folge. Leipzig, 1898.
16. Berthelot, P. *Compt. rend. 126*, 954 (1898).
17. Beattie, J. A., and Bridgeman, O. C. *Proc. Am. Acad. Arts Sci. 63*, 229 (1928).
18. Fowler, R. H. *Statistical Mechanics*. Cambridge Univ. Press, 1936.
19. Lewis, G. N., and Randall, M. *Thermodynamics*. McGraw-Hill, 1923.
20. Griffith, W. *J. Appl. Phys. 21*, 1319 (1950).
21. Rankine, W. J. M. *Trans. Roy. Soc. London A160*, 277 (1880).
22. Chapman, D. L. *Phil. Mag. 47 (5)*, 90 (1899).
23. Jouguet, E. *Compt. rend. 132*, 573 (1901).
24. Jones, H. *Proc. Roy. Soc. London A189*, 415 (1947).

Bibliography

Brason, F. W. *U.S. Bur. Mines Tech. Paper 632*, 1941.
Hottel, H. C., et al. *Trans. Am. Soc. Mech. Engrs. 70*, 667 (1948).
Huff, V. N., and Morrell, V. E. *NACA Tech. Note 2113*, 1950.

SECTION B

EXPANSION PROCESSES

DAVID ALTMAN
JAMES M. CARTER

B,1. Classification of Flow Processes. The statement of the conservation of energy for flow processes in gases where useful work other than expansion is neglected can be written in the following form:

$$dE = dq - pdV + dW_f \tag{1-1}$$

where dq is the increment of heat absorbed by the gas and dW_f the increment of frictional work done on the gas. For purposes of generality, another term may be included representing work other than expansion or friction. Since such work is not considered here, it is excluded in the following treatment in order to simplify subsequent equations. An alternate expression involving the enthalpy or heat content H can be derived through the relation

$$H = E + pV \tag{1-2}$$

or in differential form

$$dH = dE + pdV + Vdp \tag{1-3}$$

Combining Eq. 1-3 with 1-1 results in the alternate formulation

$$dH = dq + Vdp + dW_f \tag{1-4}$$

It is shown in the following treatment that either of these simple formulations of the energy equation can be utilized directly to interpret the effects of heat transfer, flow with friction, and heat release due to transitions among internal energy states and due to chemical reactions. The element of volume, however, is so chosen that uniform properties are assumed at any instant of flow within the differential element.[1] If the cross-sectional area of this element can be made as large as that of the duct (implying uniform properties throughout any cross section of flow), then simple relations will generally result relating the various thermodynamic quantities. Unless otherwise specified, it is assumed that this condition is realized for the integrated relations obtained in this section.

---

[1] For the flow of rarefied gases, it may not be possible to choose such an element because of the large mean free path. The discussion of such flows is contained in I,I and IV,E.

Corrections for nonuniform flow can be made with the relations developed from the transport properties of gases as developed in I,D.

For the complete representation of the thermodynamic parameters during a flow process, the conservation of energy equation must be combined with the conservation of momentum. For one-dimensional flow, the Bernoulli equation per unit mass of gas is

$$-\frac{dp}{\rho} = udu + dw_f \tag{1-5}$$

where w_f is the frictional work per unit mass. Noting that $\rho = m/V$, Eq. 1-5 may also be expressed as

$$-Vdp = mudu + dW_f \tag{1-5a}$$

for a gas of total mass m. Combination of Eq. 1-4 and 1-5a gives the relation between the enthalpy and the velocity,

$$dH = dq - mudu \tag{1-6}$$

Eq. 1-1 through 1-6 may be employed to treat the thermodynamics of any type of flow process. Although hydrodynamically it is convenient to classify fluid flow according to the state of motion of the fluid element, i.e. steady, irrotational, one-dimensional, etc., the present classification is made in a thermodynamic sense. Accordingly, the flow process is categorized as follows:

ADIABATIC. Any process occurring in a thermally insulated system so that $q = 0$ is an adiabatic process.

Isentropic. Any adiabatic and reversible process with $\Delta S = 0$, where S is the entropy, is an isentropic process. This category applies to systems with free flow or with work extracted, for which the internal energy states and chemical composition are either completely frozen or in equilibrium so that the composition is determined by the parameters of state.

Nonisentropic. Any adiabatic irreversible process with $\Delta S \neq 0$ is a nonisentropic process. The irreversibility may be due to many causes, some of which are nonequilibrium chemical reaction or internal energy transitions during flow, occurrence of friction, free expansion (isoenergetic, $\Delta E = 0$), and throttled flow (Joule-Thomson porous plug experiment which is isenthalpic, $\Delta H = 0$).

NONADIABATIC. Any process occurring in which heat is transferred to or from the surroundings so that $q \neq 0$ is a nonadiabatic process.

Heat transfer. Because this process must occur with a finite temperature gradient, it is necessarily irreversible.

Isothermal. This process represents the ideal case of heat transfer occurring so slowly that the temperature gradients are negligible and the temperature remains essentially constant. Although these requirements are difficult to realize in dynamic systems, the condition can be approached for low flow velocities. Such flow is usually reversible.

B,2. Thermodynamic Relations for Flow Processes.

ADIABATIC FLOW.   Adiabatic flow is the most usual type of gas flow in expansion processes occurring in practice because the duration of flow is generally so short that heat flow to or from the surroundings can be considered negligible. If the process is reversible, the flow is isentropic and exact expressions can be obtained which relate the temperature, pressure, and velocity of the gas. If friction or combustion occurs during flow, the process is nonisentropic, and relations between T and p can be obtained only if the irreversible quantities can be related, either theoretically or empirically, to the temperature, pressure, or velocity.

Isentropic flow.

1.   Free flow of nonreacting gases.   For the case of the reversible free flow where no external work is done, Eq. 1-1 can be employed directly, setting $dq = dW_f = 0$.   Therefore,

$$dE = -pdV \tag{2-1}$$

If the gases are perfect and if the heat capacity is independent of temperature, a simple integration is possible as follows (for n moles of gas):

$$C_V dT = -\frac{nRT}{V} dV \tag{2-2}$$

$$\frac{dT}{T} = -\frac{nR}{C_V}\frac{dV}{V} \tag{2-3}$$

and finally,

$$TV^{nR/C_V} = TV^{\gamma-1} = \text{const} \tag{2-4}$$

where $\gamma = C_p/C_V$ and $C_p - C_V = nR$. Alternative expressions to Eq. 2-4 can be obtained by means of the gas law $pV = nRT$. They are

$$pV^\gamma = \text{const} \tag{2-5}$$

and

$$Tp^{-nR/C_p} = Tp^{(1-\gamma)/\gamma} = \text{const} \tag{2-6}$$

The kinetic energy acquired by the gas is obtained by direct integration of Eq. 1-6 noting again that $dq = dW_f = 0$.

$$H_2 - H_1 = -\tfrac{1}{2}m(u_2^2 - u_1^2) \tag{2-7}$$

or

$$H + \tfrac{1}{2}mu^2 = \text{const} \tag{2-8}$$

for a gas which has a total mass m. Whereas Eq. 2-2 to 2-6 are based on perfect gases and constant specific heat, Eq. 2-7 and 2-8 are obtained directly from the conservation of energy and Bernoulli's equation, and hence are true for any real gas.

2.   Isentropic flow with work extracted.   If the gas does expansion work during flow and if its own kinetic energy is small relative to the work

done, then by means of Eq. 2-1 the change in internal energy is

$$E_2 - E_1 = -\int_{V_1}^{V_2} p \, dV \tag{2-9}$$

Noting the relation between p and V in Eq. 2-5, the change in E for perfect nonreacting gases is found to be

$$E_2 - E_1 = -\frac{p_2 V_2^\gamma}{1 - \gamma}[V_2^{1-\gamma} - V_1^{1-\gamma}] \tag{2-10}$$

where subscripts 1 and 2 refer to initial and final states respectively. Noting that $p_2 V_2^\gamma = p_1 V_1^\gamma = nRT_1 V_1^{\gamma-1}$ and that $C_V = nR/(\gamma - 1)$, Eq. 2-10 can be shown to reduce to

$$E_2 - E_1 = \frac{nRT_1}{1 - \gamma}\left[1 - \left(\frac{V_2}{V_1}\right)^{1-\gamma}\right]$$

$$= C_V T_1\left[\left(\frac{V_2}{V_1}\right)^{1-\gamma} - 1\right] \tag{2-10a}$$

For the case where the kinetic energy of the gases cannot be neglected (as in gun barrels), a more detailed treatment is necessary which takes into account the pressure and velocity distribution of the gas in the duct, [1]. For such cases, Eq. 2-10a is usually taken as a first approximation. In evaluating various powder grains in a given gun where V_2/V_1 is fixed, the quantity nRT_1, known as the impetus, is found to be a very useful parameter.

3. Effect of equilibrium chemical reaction and internal energy state transitions. For calculations with gases for which the specific heat varies markedly with temperature (resulting from internal energy state changes), or in systems with shifting chemical equilibria, Eq. 2-4, 2-5, and 2-6 are approximate, subject to the choice of a proper average C_p or γ for the process. Accurate calculations in these cases are made by use of the statement for isentropic flow:

$$\Delta S = 0 \tag{2-11}$$

with

$$S = \sum n_i (S_{298}^0)_i + \sum n_i \int_{298}^{T} C_{p_i} d(\ln T) - nR \ln p - \sum n_i R \ln \frac{n_i}{n} \tag{2-12}$$

For systems involving equilibrium chemical reaction during flow, the composition of the system, given by the n_i values, is determined at a trial temperature for a given pressure. The entropy S, at the temperature, may now be calculated from Eq. 2-12 and compared with the entropy in the initial state. Interpolation between several trial temperatures is then made to yield that value of S satisfying Eq. 2-11.

For shifting chemical equilibria, the enthalpy H is defined so that it contains the heat of chemical reaction resulting from composition changes. This definition is (for perfect gases)

$$H = \sum n_i [\Delta H_{f(298)}]_i + \sum n_i \int_{298}^{T} C_{p_i} dT \qquad (2\text{-}13)$$

where $[\Delta H_{f(298)}]_i$ is the heat of formation per mole of species i from the elements at the base temperature 298°K. For constant composition, $(d/dT)\sum n_i[\Delta H_{f(298)}]_i = 0$ and so the enthalpy of each species is simply taken relative to its own value at the base temperature.[2] Fairly complete tabulations of the standard entropy

$$S = S_{298}^{0} + \int_{298}^{T} C_p d (\ln T)$$

the relative enthalpy

$$H = \int_{298}^{T} C_p dT$$

and the standard heat of formation $\Delta H_{f(298)}$ are given by the National Bureau of Standards [2].

In systems where the chemical composition is frozen but where internal energy state transitions occur, causing the specific heat to vary with the temperature, it is found that the Eq. 2-4 to 2-8 can be applied directly with good accuracy if an appropriate average $\bar{C}_p$ is employed. Several methods for choosing $\bar{C}_p$ have been considered [3,4], a simple and fairly accurate one being $\bar{C}_p = (H_1 - H_2)/(T_1 - T_2)$ where T_2 is a tentative final temperature. This value of $\bar{C}_p$ is very close to that calculated at the temperature $\frac{1}{2}(T_1 + T_2)$. These approximations are good because the enthalpy change is not highly sensitive to small variations in C_p or γ.

If analytic functions for the specific heat are available, direct integration of the energy equation is possible. Thus if the heat capacity of the gas can be expressed as

$$C_p = a + bT + cT^2 \qquad (2\text{-}14)$$

the T, p relation is found to be

$$\frac{T_2}{T_1} = \left(\frac{p_2}{p_1}\right)^{\frac{nR}{a}} e^{-\left[\frac{b}{a}(T_1 - T_2) + \frac{c}{2a}(T_1^2 - T_2^2)\right]} \qquad (2\text{-}15)$$

and the enthalpy change between T_2 and T_1 is

$$\Delta H_{T_1}^{T_1} = a(T_1 - T_2) + \frac{b}{2}(T_1^2 - T_2^2) + \frac{c}{3}(T_1^3 - T_2^3) \qquad (2\text{-}16)$$

Nonisentropic flow: effects of friction and combustion. The most common causes for nonisentropic flow in adiabatic systems are the occurrence of frictional effects and combustion. Although the viscosity of the gas can generally be neglected, the frictional effects at the wall may become appreciable for very small ducts.

[2] The choice of the base temperature at 298°K, 0°K, or any other temperature is entirely arbitrary and would yield identical results.

1. **Friction.** It will be assumed that (1) the velocity is uniform across the section of the duct and (2) the viscosity is concentrated at the wall with the force determined empirically. Reference to Fig. B,2a shows that the pressure force acting on an elementary volume of length dx is

$$-A\,dp = A\rho u\,du + f l\,dx \tag{2-17}$$

where f is the frictional force per unit area and l the periphery. Dividing through by $A\rho$ results in

$$\frac{-dp}{\rho} = u\,du + \frac{f l\,dx}{A\rho} \tag{2-18}$$

Eq. 2-18 is essentially the same as Eq. 1-5 where the term dw_f is found to be equal to $f l\,dx / A\rho$ which represents the frictional heat developed per

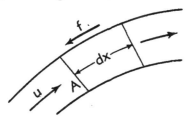

Fig. B,2a.

unit mass of gas as a result of the frictional work. The velocity can be determined directly from Eq. 1-6 where $dq = 0$ for adiabatic flow. The result is

$$dH + m\,d\left(\frac{u^2}{2}\right) = 0 \tag{2-19}$$

Eq. 2-19 is significant in that it shows the velocity to be given directly by the enthalpy drop as in frictionless flow. The effect of friction, therefore, must be only to decrease the enthalpy drop between two pressure levels. Eq. 2-19 cannot yet be employed to determine the flow velocity until the T, p relation has been found which permits the enthalpy drop between two pressure levels to be calculated. This relation can be obtained from Eq. 1-4 setting $dq = 0$ and $dW_f = V f l\,dx / A$. The result is

$$dH = V\,dp + \frac{V f l\,dx}{A} \tag{2-20}$$

Exact integration of this equation cannot be made unless the dependence of f on p and x is known. However, since the frictional work is completely manifested by heat, let $V f l\,dx / A = dq_i$. For a perfect gas with constant specific heat, Eq. 2-20 can now be rewritten as

$$C_p \frac{dT}{T} = nR \frac{dp}{p} + \frac{dq_i}{T} \tag{2-21}$$

in which dq_i/T can be identified with dS_i, the entropy gained through frictional heating. Integration from T_1 and p_1 to T and p yields

$$\frac{T}{T_1} = \left(\frac{p}{p_1}\right)^{\frac{nR}{C_p}} e^{\frac{\Delta S_i}{C_p}} \tag{2-22}$$

with $T > T_0$ for all frictional processes. Fig. B,2b and B,2c show the p, T and H, S diagrams for isentropic flow and flow with friction. The

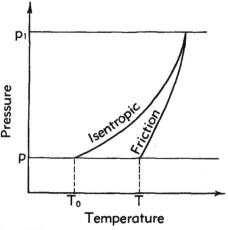

Fig. B,2b. *pT* diagram for isentropic flow and flow with friction.

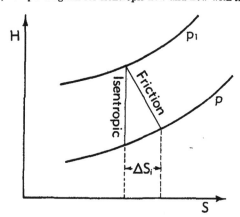

Fig. B,2c. *HS* diagram for isentropic flow and flow with friction.

irreversible entropy gained, ΔS_i, can be evaluated from Eq. 2-23 in terms of T and T_0 to yield

$$\Delta S_i = C_p \ln\left(\frac{T}{T_0}\right) \tag{2-23}$$

The adiabatic efficiency η, defined as the ratio of kinetic energy with friction to that without friction, is given by the expression

$$\eta = \frac{T_1 - T}{T_1 - T_0} \qquad (2\text{-}24)$$

The expansion law may be obtained from Eq. 2-24 by multiplying both sides of the equation by $(T_1 - T_0)/T_1$ and identifying T_0/T_1 as $(p/p_1)^{nR/C_p}$. The result is[3]

$$\frac{T}{T_1} = \eta\left(\frac{p}{p_1}\right)^{\frac{nR}{C_p}} + (1 - \eta) \qquad (2\text{-}25)$$

This law is sometimes represented with a polytropic expansion constant [5,6] which is an approximate form of Eq. 2-25. Since $nR/C_p < 1$, Eq. 2-25 may be expanded to yield

$$\frac{T}{T_1} = 1 - \eta + \eta + \frac{\eta nR}{C_p}\ln\left(\frac{p_1}{p}\right) + \cdots$$

$$\cong 1 + \frac{\eta nR}{C_p}\ln\frac{p}{p_1} \qquad (2\text{-}26)$$

Again, since $\eta nR/C_p < 1$, it will be observed that Eq. 2-26 gives the first two terms of the expansion for $(p/p_1)^{\eta nR/C_p}$ and so there finally results

$$\frac{T}{T_1} \cong \left(\frac{p}{p_1}\right)^{\frac{\eta nR}{C_p}} \qquad (2\text{-}27)$$

which may also be expressed in the equivalent form (by use of the perfect gas law):

$$pV^k \cong \text{const} \qquad (2\text{-}28)$$

where $k = \gamma/[\gamma - \eta(\gamma - 1)]$ is the polytropic constant.

These polytropic expansion laws have also been derived [5,6] by means of the equation $dq_i = C_p(dT - dT_0)$ and Eq. 2-21. However, it should be noted that this relation for dq_i is only approximate since the difference in the expansion work terms $\int V dp$ for the viscous and inviscid gas between the fixed pressure levels has been neglected.

2. Combustion and nonequilibrium reaction during flow. The case of combustion during adiabatic flow, unlike that of equilibrium chemical reassociation, is nonisentropic because the chemical composition is not a unique function of the equation of state of the gas. The energy equation governing such flow is given by Eq. 1-4 with $dq = dW_f = 0$ for adiabatic frictionless flow.[4] The heat content H is given by Eq. 2-13 presented

---

[3] Strictly speaking η may not be constant during flow, but for approximate calculations the assumption of constancy is fairly good [5,6].

[4] Some authors prefer to treat combustion during flow as nonadiabatic, identifying the chemical heat with q [7]. This treatment will yield equivalent results if the reference state for H is defined as is done for a nonreactive gas.

earlier:

$$H = \sum n_i [\Delta H_{f(298)}]_i + \sum n_i \int_{298}^{T} C_{p_i} dT \qquad (2\text{-}13)$$

where $[\Delta H_{f(298)}]_i$ is the heat of formation of the ith species at 298°K, the reference state being the elements at 298°K. For perfect gases, H is a function only of T and n_i, and taking the differential of H gives

$$dH = \sum [\Delta H_{f(298)}]_i dn_i + \sum dn_i \int_{298}^{T} C_{p_i} dT + \sum n_i C_{p_i} dT$$

$$= \sum [\Delta H_i]_i dn_i + \sum n_i C_{p_i} dT \qquad (2\text{-}29)$$

where $[\Delta H_i]_i$ is the heat of formation at the temperature of the gas. The energy equation may now be written as

$$\sum [\Delta H_i]_i \frac{dn_i}{dT} dT + \sum n_i C_{p_i} dT = \frac{nRT}{p} dp \qquad (2\text{-}30)$$

where $n = \sum n_i$. Dividing through by nT and denoting $C_p = \sum n_i C_{p_i}$ as the average specific heat of gas results in

$$\sum \frac{[\Delta H_i]_i}{n} \frac{dn_i}{dT} \frac{dT}{T} + \frac{C_p}{n} \frac{dT}{T} = R \frac{dp}{p} \qquad (2\text{-}31)$$

This equation is not integrable since n_i cannot be expressed as a unique function of the state of the gas because of its dependence on time. For a fixed geometry, however, the parameters T and p can be related to the velocity and hence to the time. If the reaction rate constants of the various combustion reactions are known, then expressing dn_i/dT as $(dn_i/dt)(dt/dT)$ will permit solution of Eq. 2-31 because these derivatives can be expressed analytically. This method of treatment has been applied to the dissociation of NO during flow through a nozzle [8].

Approximate solutions to Eq. 2-31 can be obtained if one knows the total chemical heat released and assumes the heat to be generated uniformly. The quantity $\sum([\Delta H_i]_i/n)(dn_i/dT)$ is replaced by $\Delta Q/\bar{n}(T_1 - T_2')$ where ΔQ is the total chemical heat of reaction, $\bar{n}$ is the average number of moles of gas considered, and T_2' is a tentative value of the final temperature. It will be observed that the quantity $\Delta Q/(T_1 - T_2')$ has the dimensions of heat capacity and can be denoted by C_p'. Eq. 2-31 now integrates to

$$T = \text{const} \times p^{\frac{\bar{n}R}{C_p + C_p'}} = \text{const} \times p^{\frac{\bar{\gamma}-1}{\bar{\gamma}}} \qquad (2\text{-}32)$$

with an apparent

$$\bar{\gamma} = \frac{C_p + C_p'}{C_V + C_V'}$$

The low value of the apparent $\bar{\gamma}$ will lead to higher stream temperatures as shown in Fig. B,2d.

3. Throttled flow: isenthalpic. Steady state adiabatic flow from a high pressure region p_1 through a valve or porous plug to a low pressure region p_2 occurs at constant enthalpy. This result was first obtained in the classical experiments of Joule and Thomson in their famous porous plug experiments.

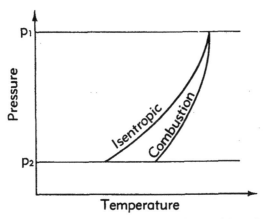

Fig. B,2d. pT diagram for isentropic flow and flow with combustion.

Employing the integrated form of Eq. 1-1 in which friction is neglected, the energy change per mole of gas in going from p_1 to p_2 is given by the equation

$$E_2 - E_1 = (p_1V_1 - p_2V_2) \tag{2-33}$$

where the net work produced by the gas is the work done by the gas at p_1, i.e. p_1V_1, minus that done on the gas at p_2, i.e. p_2V_2. Noting that the enthalpy H is given by the equation, $H = E + pV$, it is seen from Eq. 2-33 that the quantity

$$H_2 = E_2 + p_2V_2 = E_1 + p_1V_1 = H_1 \tag{2-34}$$

remains constant. If the gas is perfect so that H is a function only of T, the process is also isothermal. If the gas is imperfect, or if chemical equilibria exist which are pressure dependent, the process is still isenthalpic but not isothermal. This principle has found application in gas generation devices where the gas is formed at a high pressure and does work at a lower pressure.

4. Free expansion: isoenergetic. Free expansion of a gas occurs in the transfer of the gas from a higher to a lower pressure level without the appearance of external work. If the system is adiabatic and frictionless, it is obvious from Eq. 1-1 that the over-all process occurs at constant internal energy for the system, i.e. $\Delta E = 0$. For imperfect gases, E is a

function of both T and p and so the energy change associated with the change in p must be exactly compensated by a change in T. At low temperatures, this effect results in a cooling of the gas. This phenomenon is made use of in one method for the liquefaction of low boiling-point gases.

NONADIABATIC FLOW. In any real system, the assumption of adiabatic behavior can only be employed when the heat transferred through the walls is small with respect to heat effects produced in the gas. Nonadiabatic flow, on the other hand, is favored when (1) wall surface per unit mass of gas is large, (2) thermal conductivity of the gas and wall material is high, (3) wall temperatures are widely different from that of the gas, and (4) the flow velocity is low. The degree of irreversibility of the heat transfer process may be measured by the difference in temperature between the wall and the gas. In a thermodynamic sense, this irreversibility resulting from the transfer of heat is given by the entropy increase of the entire system:

$$\Delta S_1 + \Delta S_2 = \frac{q}{T_1} - \frac{q}{T_2} = q\frac{\Delta T}{T_1 T_2} > 0 \qquad (2\text{-}35)$$

where T_1 and T_2 are the lower and higher temperatures, respectively, and ΔS_1 and ΔS_2 the corresponding entropy changes. As $\Delta T \to 0$ in Eq. 2-35, the process becomes isothermal and reversible. The conditions for this behavior are discussed under *Isothermal expansion* on page 39.

Heat transfer. It is assumed that the heat is transferred from the walls into the gas stream normal to the flow velocity and that the heat flow along the flow path is negligible. It is obvious that in any real case, a temperature gradient will necessarily become established normal to the flow because of the limitation of thermal conductivity.

The treatment given here will be limited to the thermodynamic behavior of a small element of fluid in which average properties can be chosen. The relations so obtained can be applied quite successfully to low velocity turbulent streams with large l/D ratios. Application may also be made to systems where the heat transferred, q, is much less than the enthalpy potential $C_p T$. It can be shown that subject to this limitation, specific consideration of the cross-sectional temperature contour will give rise only to second order corrections to the flow parameters. This fact is of practical significance since it permits the direct application of the following treatment to heat transfer in supersonic nozzles where $q \ll C_p T$ and where the cross-sectional contour cannot be uniform.

The energy equation (1-4) can be applied by setting $dW_f = 0$. Division by $C_p T$ results in

$$\frac{dT}{T} = \frac{dq}{C_p T} + \frac{nRdp}{C_p p} \qquad (2\text{-}36)$$

Direct integration from T_1 and p_1 to T and p yields the equation

$$\frac{T}{T_1} = \left(\frac{p}{p_1}\right)^{\frac{nR}{C_p}} e^{\frac{\Delta S}{C_p}} \tag{2-37}$$

which is identical to that which was given for flow with friction, Eq. 2-22, showing that thermodynamically the effect of friction on the T, p relation can be treated simply as a heat transfer process. One significant point of difference must be noted, however, namely, that the transfer of a definite quantity of heat between two fixed pressure levels will in general give a different T, p relation for the two processes. This arises from the fact that the temperature rate of transfer may be different for the two flows resulting in different values of ΔS.

The flow velocity and kinetic energy of the stream is obtained by integrating Eq. 1-6.

$$H_2 - H_1 + \tfrac{1}{2}m(u_2^2 - u_1^2) = q \tag{2-38}$$

where q represents the quantity of heat transferred to the gas between T_1 and T_2.

Eq. 2-37 and 2-38 cannot be employed for quantitative calculations unless the entropy change ΔS can be evaluated. Since q is not a function of state, such an evaluation can be made only when the geometry and heat transfer rates are known for the system. In certain applications with fixed geometry, however, simple approximate relations between q, S, and T can be obtained. One such case occurs in the heat transfer through rocket nozzles.

1. Approximate T, p relation for flow through rocket nozzles with heat transfer. Measurements of heat transfer rates through sectional nozzles [5] show that the maximum rate per unit area of nozzle occurs at the throat. However, per unit mass of gas, the maximum occurs somewhat upstream of the throat. Since the throat temperature given by $T_t = (2/(\gamma + 1))T_o$, is generally close to the chamber temperature T_o, it is obvious that an average temperature defined through the relation

$$\int_{T_o}^{T_o} \frac{dq}{T} = \frac{q}{T_{av}}$$

will be close to T_t. The fact that heat transferred downstream of the throat at lower temperatures will cause a greater change in entropy is partially compensated by the fact that the maximum heat transferred per unit mass lies upstream of the throat [5]. These facts appear to justify the approximate relation, where q_N is the heat transferred out of the gas during flow through the nozzle,

$$\Delta S \cong \frac{q_N}{T_t} = -\frac{(\gamma + 1)}{2T_o} q_N \tag{2-39}$$

for nozzles of conventional design (see XII,G).

Combination of Eq. 2-37 and 2-39 will now yield the following equation between T, p and q_N:

$$\frac{T}{T_o} = \left(\frac{p}{p_o}\right)^{\frac{nR}{C_p}} e^{-\frac{(\gamma+1)q_N}{2C_pT_o}} \tag{2-40}$$

or in terms of γ:

$$\frac{T}{T_o} = \left(\frac{p}{p_o}\right)^{\frac{\gamma-1}{\gamma}} e^{-\frac{(\gamma^2-1)}{2\gamma}\frac{q_N}{nRT_o}} \tag{2-40a}$$

Subject to the earlier assumption of $q_N \ll C_pT_o$, the exponential may be expanded to yield

$$\frac{T}{T_o} = \left(\frac{p}{p_o}\right)^{\frac{\gamma-1}{\gamma}}\left[1 - \left(\frac{\gamma+1}{2}\right)\frac{q_N}{C_pT_o}\right] \tag{2-41}$$

Eq. 2-38 and 2-40 establish the relations necessary for correcting exhaust velocities in rockets for nozzle heat transfer.

2. Exhaust velocity corrections resulting from nozzle heat transfer. The exhaust velocity in an adiabatic nozzle resulting from expansion of the gases at rest at T_o and p_o to T_e and p_e is given by the equation

$$\frac{1}{2}mu_e^2 = H_o - H_e = C_p(T_o - T_e) = C_pT_o\left(1 - \frac{T_e}{T_o}\right) \tag{2-42}$$

Denoting the quantity

$$1 - \frac{T_e}{T_o} = 1 - \left(\frac{p_e}{p_o}\right)^{\frac{\gamma-1}{\gamma}}$$

by η_i, the thermodynamic efficiency, Eq. 2-42 may be written simply as

$$\tfrac{1}{2}mu_e^2 = \eta_i C_pT_o \tag{2-42a}$$

It will be of interest now to compare the exhaust velocity in a water-cooled nozzle u_w with that in a regeneratively cooled nozzle u_r where the heat transferred, q_N, is fed back into the combustion zone at T_o and p_o. For the case where the heat transferred to the surroundings is lost (water cooling), the kinetic energy equation is given by Eq. 2-38. For expansion nozzles with small divergence angles or large expansion ratios, T_{av} should be chosen somewhat less than T_t.

$$\frac{1}{2}mu_w^2 = C_pT_o\left(1 - \frac{T_e}{T_o}\right) - q_N \tag{2-43}$$

Substituting for T_e/T_o from Eq. 2-41, there finally results

$$\frac{1}{2}mu_w^2 = \eta_i C_pT_o\left[1 + \left(\frac{\gamma+1}{2}\right)\frac{q_n}{C_pT_o}\left(\frac{1-\eta_i}{\eta_i}\right)\right] - q_N \tag{2-44}$$

When the heat transferred through the nozzle, q_N, is fed back into the combustion chamber, the chamber temperature is increased by the factor $(1 + q_N/C_pT_c)$. Aside from this change, the remainder of the process is the same as that for water cooling.[5] Hence, there results the equation

$$\frac{1}{2} mu_r^2 = \eta_i C_p T_o \left(1 + \frac{q_N}{C_pT_o}\right)\left[1 + \left(\frac{\gamma + 1}{2}\right)\frac{q_N}{C_pT_o}\left(\frac{1 - \eta_i}{\eta_i}\right)\right] - q_N \tag{2-45}$$

The exhaust velocity ratios for these three systems may be formulated as follows:

$$\frac{u_w}{u_o} = 1 - \frac{1}{2\eta_i}[1 - \tfrac{1}{2}(\gamma + 1)(1 - \eta_i)]\frac{q_N}{C_pT_o} \tag{2-46}$$

$$\frac{u_r}{u_o} = 1 + \frac{1}{4\eta_i}[(1 - \eta_i)(\gamma - 1)]\frac{q_N}{C_pT_o} \tag{2-47}$$

Eq. 2-47 yields an interesting result for it indicates a means of obtaining higher than adiabatic performance through regenerative heat transfer. This improvement in performance results from the fact that heat which is abstracted at a low pressure level is put back into the system at a higher pressure level. For very large pressure ratios, however, this effect is diminished since as $u_r/u_o \to 1$, $\eta_i \to 1$. The result in this limiting case is a direct consequence of the conservation of energy since at $\eta_i = 1$ the entire enthalpy is converted into kinetic energy in the adiabatic nozzle and obviously no scheme can cause an improvement.

Isothermal expansion. In practice, expansion processes which are isothermal are generally fairly difficult to realize because of limitations of thermal conductance; nevertheless, it is instructive to analyze the thermodynamic behavior of such flow for comparison with the more usual adiabatic case. These equations have found practical utility in systems with heat transfer to the gas from an external thermostat at the stagnation temperature of the stream (gas pressurization systems). Under these conditions, the isothermal expansion provides an upper limit for the efficiency of the expansion process. Isothermal expansion processes in rocket nozzles have been considered theoretically [9] but because of the short residence time, practical application does not seem feasible.

For isothermal expansion in a frictionless stream, the heat transferred to the gas between two pressure levels is uniquely determined and hence its transfer is reversible. This permits immediate identification of the entropy increment accordingly:

$$dq = TdS \tag{2-48}$$

The energy equation may be applied to the process in the following form

[5] For the case where q_N is an appreciable fraction of C_pT_o, the value of q_N in the water-cooled and regeneratively cooled nozzles may not be the same because of the higher temperature in the regenerative case. Second order corrections of this sort are neglected in this treatment.

$(dW_f = 0)$:

$$\left(\frac{\partial E}{\partial p}\right)_T dp = dq - pdV = TdS - pdV \qquad (2\text{-}49)$$

where the term $(\partial E/\partial p)_T$ contains the energy of imperfection. For perfect gases, this term is zero and direct integration of Eq. 2-49 yields for the heat absorbed by the gas

$$q = nRT \ln\left(\frac{V_2}{V_1}\right) = T(S_2 - S_1) \qquad (2\text{-}50)$$

The equation for the velocity is obtained from Eq. 1-6 and 2-48 to give

$$dH - TdS + mudu = 0 \qquad (2\text{-}51)$$

Since $F = H - TS$, the first two terms on the left may be identified with dF for an isothermal process. Integration finally yields

$$F + \tfrac{1}{2}mu^2 = \text{const} \qquad (2\text{-}52)$$

which may be contrasted with Eq. 2-8 for adiabatic flow. The free energy change is given in terms of the pressure gradient of F, i.e. $(\partial F/\partial p)_T = V$ which integrates to

$$F_2 - F_1 = \int_{p_1}^{p_2} Vdp \qquad (2\text{-}53)$$

The kinetic energy increment in terms of the pressure drop is obtained from Eq. 2-52 and 2-53 to yield

$$\tfrac{1}{2}m(u_2^2 - u_1^2) = \int_{p_1}^{p_2} Vdp \qquad (2\text{-}54)$$

and for perfect gases, this reduces to

$$\frac{1}{2}m(u_2^2 - u_1^2) = nRT \ln\left(\frac{p_1}{p_2}\right) \qquad (2\text{-}55)$$

B,3. Determination of Performance Parameters for Isentropic Flow.

Exhaust velocity and specific impulse. As given in Art. 1 and 2 above, the exhaust velocity or specific impulse for isentropic expansion is simply obtained from the initial and final states. For a gas with negligible initial velocity, the exhaust velocity is accurately given by the formula (cf. Eq. 2-42):

$$u_e = \sqrt{\frac{2\Delta H}{m}} \qquad (3\text{-}1)$$

If the gases are perfect and chemical reactions during flow are negligible, Eq. 3-1 can be expressed in any one of the following forms where T_e is the chamber temperature and C_p is the heat capacity (assumed constant):

$$u_e = \sqrt{\frac{2C_p T_c}{m} \left[1 - \left(\frac{p_e}{p_c} \right)^{\frac{nR}{C_p}} \right]} \tag{3-1a}$$

or

$$u_e = \sqrt{\frac{2\gamma R T_c}{(\gamma - 1)\mathfrak{M}} \left[1 - \left(\frac{p_e}{p_c} \right)^{\frac{\gamma - 1}{\gamma}} \right]} \tag{3-1b}$$

The enthalpy of the combustion gases at combustion conditions can be assumed accurately known from the composition and thermochemical data. The enthalpy at exhaust conditions may be calculated more or less accurately for various conditions of flow as outlined in Art. 2.

The computed exhaust velocity may be compared with the experimental thrust by the formula:[6]

$$F = \lambda u_e \dot{m} + (p_e - p_0) A_e \tag{3-2}$$

or

$$c = \lambda u_e = \frac{F - (p_e - p_0) A_e}{\dot{m}} \tag{3-3}$$

where $\dot{m} = dm/dt$ is the mass rate of flow and p_0 the ambient pressure.

In most cases the expansion nozzle is so proportioned that p_e is very nearly equal to p_0. For this condition the pressure term drops out, and the effective exhaust velocity is simply the ratio of thrust to mass rate of flow.

The parameter specific impulse, I_{sp}, is defined as the impulse per unit weight of propellant, Ft/w, or the thrust per unit weight rate of flow, $F/\dot{w}$. Both definitions are equivalent for steady state conditions. By means of Eq. 3-2, this definition results in the equation:

$$I_{sp} = \frac{c\dot{m} + (p_e - p_0) A_e}{\dot{w}} = \frac{c}{g} + \frac{(p_e - p_0) A_e}{\dot{w}} \tag{3-4}$$

For the nozzle so designed that $p_e = p_0$, I_{sp} is simply related to c as follows:

$$I_{sp} = \frac{c}{g} \tag{3-4a}$$

For most propellants for which combustion is known to be efficient, the experimental velocity as determined by Eq. 3-3 will be some 90 per cent or more of the value computed by Eq. 3-1. In some cases when heat loss to the combustion chamber and nozzle are known, and when other experimental conditions are ideal, the experimental velocity may be as high as 98 per cent of the computed value.

Thus the enthalpy drop converted to kinetic energy will usually range from 80 to 95 per cent of that computed by thermodynamic methods.

[6] For converging-diverging nozzles, only the axial component of the velocity will produce a net thrust. In comparing experimental and theoretical thrust values, therefore, one must apply a geometric correction λ to u_{eff} [14], which has the value $\lambda = \frac{1}{2}(1 + \cos \alpha)$ where α is the half angle in the divergence section. The "effective velocity" λu_{eff} is sometimes designated by c.

Among the factors which contribute to the loss are (1) loss of heat to the walls of combustion chamber and nozzle, (2) friction in the nozzle, and (3) angle of divergence of the nozzle. These are inherent in any rocket motor. Other less apparent factors are incomplete combustion, combustion continuing in the nozzle, uncertainty as to maintenance of equilibrium during expansion, and shock effects which may arise as soon as sonic velocity is reached.

Characteristic velocity. Although over-all performance as given by the exhaust velocity or specific impulse is easily correlated with theoretical data, the same is not true for some other performance parameters. In any experimental rocket motor test, the following quantities are ordinarily measured or known:

1. Thrust F
2. Combustion pressure p_c
3. Exit pressure p_e (usually made equal to p_0)[7]
4. Nozzle throat area A_t
5. Nozzle exit area A_e
6. Weight rate of propellant flow $\dot{w} = dw/dt$
7. Propellant composition

If exhaust and ambient pressures are equal, or if their difference is measured, the experimental exhaust velocity is determined from items (1) and (6). It may also be computed under ideal conditions from (2), (3), and (7).

In some test work, it is convenient to separate, as far as possible, the combustion and expansion processes. For a given propellant combination, the combustion process should determine the intrinsic capability to generate kinetic energy; the expansion process determines the effectiveness of use of the heat (or work) originally present.

For this reason, use is made of two experimental quantities, known as the "characteristic velocity" (c^*) and the "nozzle thrust coefficient" (C_F). These are defined by the following relations:

$$c^* = \frac{p_c A_t}{\dot{m}} \tag{3-5}$$

and

$$C_F = \frac{F}{p_c A_t} \tag{3-6}$$

It is obvious that the product of c^* and C_F is the effective exhaust velocity which equals c in Eq. 3-3 when $p_e = p_0$. The "characteristic velocity" is then the reciprocal of the mass rate of flow per unit nozzle throat area and per unit pressure in the combustion chamber. The "nozzle thrust coefficient" is the thrust per unit throat area and per unit pressure in the combustion chamber.

[7] For flight conditions with varying p_0, p_e is obtained from the area ratio A_e/A_t.

The theoretical expression for c^* is obtained from the definition, Eq. 3-5, and the equation of continuity

$$\rho u A = \dot{m} \tag{3-7}$$

where ρ is the mass density. These two equations yield (the subscript t applies to conditions at the throat)

$$c^* = \frac{p_o}{\rho_t u_t} \tag{3-8}$$

The throat is defined as that region in the nozzle with minimum cross section. From Eq. 3-7 it is seen that this condition occurs when the product ρu is a maximum. For systems where the gases are extensively dissociated and chemical equilibrium assumed, the accurate way of calculating c^*, therefore, is to determine the maximum value of ρu by calculating this quantity at several assumed pressure levels near the throat. Equations for calculating these quantities can be found in Art. 2. Once the maximum value of $\rho u = \rho_t u_t$ has been found, Eq. 3-8 can be applied directly.

The method for calculating c^* as outlined above can be very laborious because of the many equilibrium composition calculations that must be made at the various trial pressures (or temperatures) in the vicinity of the throat. For this reason, many approximate relations have been derived which have been used almost exclusively up to this time. These equations depend upon the fact that the relations for nondissociated gases with constant specific heats can be employed for dissociated systems if a proper average value of γ or C_p is chosen (see Art. 2).

Let γ_t be the average value of γ in the vicinity of the throat and γ' be the average value between the chamber and the throat. Under the assumption that the velocity in the nozzle throat is sonic under local conditions and that the perfect gas laws hold, the velocity at the throat is given by

$$u_t = \sqrt{\left(\frac{dp}{d\rho}\right)_s} = \sqrt{\frac{\gamma_t R T_t}{\mathfrak{M}_t}} \tag{3-9}$$

since $p = \text{const} \times \rho^\gamma$ according to Eq. 2-5. The equation of state at the throat is $p_t \mathfrak{M}_t = \rho_t R T_t$ which can be incorporated with Eq. 3-9 into Eq. 3-8 to give for c^* the following:

$$c^* = \frac{p_o}{p_t} \sqrt{\frac{R T_t}{\gamma_t \mathfrak{M}_t}} = \left(\frac{p_o}{p_t}\right)\left(\frac{T_t}{T_o}\right)^{\frac{1}{2}} \sqrt{\frac{R T_o}{\gamma_t \mathfrak{M}_t}} \tag{3-10}$$

The critical pressure and temperature ratios are further given by the expression [15]:

$$\frac{T_t}{T_o} = \frac{2}{\gamma' + 1} \quad \text{and} \quad \frac{p_t}{p_o} = \left(\frac{2}{\gamma' + 1}\right)^{\frac{\gamma'}{\gamma' - 1}} \tag{3-11}$$

Eq. 3-10 and 3-11 give finally

$$c^* = \sqrt{\frac{RT_c}{\gamma_t \mathfrak{M}_t} \left(\frac{\gamma' + 1}{2}\right)^{\frac{\gamma'+1}{\gamma'-1}}} \qquad (3\text{-}12)$$

or

$$c^* = \frac{\sqrt{RT_c/\gamma_t \mathfrak{M}_t}}{\left(\dfrac{2}{\gamma' + 1}\right)^{\frac{\gamma'+1}{2(\gamma'-1)}}} \qquad (3\text{-}12a)$$

For a nonreacting gas with constant specific heat, identifying marks on γ and $\mathfrak{M}$ may be removed and Eq. 3-12 is accurate.

Eq. 3-12 or 3-12a is still rather cumbersome for the chemically reacting gas because of the necessity of evaluating $\mathfrak{M}_t$, γ_t, and γ'. On the other hand, when performance calculations are normally made for rocket propellants, the quantities $\mathfrak{M}$ and γ are known in the chamber and in the exhaust. Let $\mathfrak{M}_c$ be the average molecular weight in the chamber and $\overline{\overline{\mathfrak{M}}}$ be the average during the expansion with a corresponding definition for γ_c and $\bar{\gamma}$. Employing these definitions, various investigators have found the following equations quite reliable and useful. The quantity $\sqrt{\gamma}\,[2/(\gamma + 1)]^{(\gamma+1)/2(\gamma-1)}$ has been replaced by $\Gamma(\gamma)$, values for which have been tabulated [15].

$$c_c^* = \frac{\sqrt{RT_c/\mathfrak{M}_c}}{\Gamma(\gamma_c)} \qquad (3\text{-}12b)$$

$$\overline{c^*} = \frac{\sqrt{RT_c/\overline{\overline{\mathfrak{M}}}}}{\Gamma(\bar{\gamma})} \qquad (3\text{-}12c)$$

The c^* equations (3-12a, 3-12b, and 3-12c) generally do not differ by more than 2 per cent for moderately dissociated systems, with c_c^* giving the maximum value in all cases.

From the nature of the formula for c^*, it is apparent that for nondissociated gases and in the absence of such complicating effects as heat transfer, friction, shock effects, boundary layers, etc., the value of c^* is a function only of propellant composition and combustion efficiency since only these quantities can cause a variation in T_c, $\mathfrak{M}$, and γ. For this reason, it is frequently stated that c^* is independent of expansion conditions. For chemically dissociated systems, however, the chamber pressure will influence the state of chemical equilibrium and hence the values of T_c, $\mathfrak{M}$, and γ during expansion causing c^* to be indirectly a function of pressure. Because of the rate effect in adjusting to chemical equilibria during flow, c^* will also be a function of nozzle length in the contraction section as this length affects the residence time. Generally speaking, c^* can be considered to be affected by conditions up to the throat but independent of conditions downstream of that point for uniform flow.

Thrust coefficient. The thrust coefficient C_F is defined through the relation in Eq. 3-6, and in terms of thermodynamic quantities is given by the expression

$$C_F = \frac{\lambda u_e \dot{m} + (p_e - p_0) A_e}{p_e A_t}$$

$$= \frac{c}{c^*} + \left(\frac{p_e - p_0}{p_e}\right) \frac{A_e}{A_t}$$

(3-13)

For $\lambda = 1$ and $p_e = p_0$, $C_F = C_F^0$ and is given by

$$C_F^0 = \frac{u_e}{c^*}$$

(3-14)

These equations show that unlike c^*, C_F is a function of nozzle geometry downstream of the throat, and of exit pressure.

For a perfect nonreacting gas with constant specific heat, C_F^0 is given by the expression:

$$C_F^0 = \sqrt{\frac{2\gamma^2}{\gamma - 1} \left(\frac{2}{\gamma + 1}\right)^{\frac{\gamma+1}{\gamma-1}} \left[1 - \left(\frac{p_e}{p_0}\right)^{\frac{\gamma-1}{\gamma}}\right]}$$

(3-15)

which was obtained from Eq. 3-1b, 3-12c, and 3-14. For a real gas with variable specific heat, an equation for C_F^0 similar to that given for c^* in Eq. 3-12 can be formulated with specified γ values. One such equation is

$$C_F^0 = \sqrt{\frac{2\bar{\gamma}\gamma_t}{\bar{\gamma} - 1} \left(\frac{2}{\gamma' + 1}\right)^{\frac{\gamma'+1}{\gamma'-1}} \left[1 - \left(\frac{p_e}{p_0}\right)^{\frac{\bar{\gamma}-1}{\bar{\gamma}}}\right]}$$

(3-16)

where γ', $\bar{\gamma}$, and γ_t are as defined previously (cf. Eq. 3-12, 3-12a, and 3-12c).

B,4. Nonequilibrium Effects.

Incomplete combustion. Jet engines are so designed that combustion reactions are essentially completed in the chamber prior to expansion through the nozzle. The size of the combustion chamber is determined not only by the mass rate of flow or thrust of the engine but also by the rate at which the combustion reaction proceeds. Under the assumption that the combustibles flow uniformly down the chamber without the occurrence of large eddy currents or stagnation regions (the flow, however, need not be laminar), the completeness of combustion may be measured in terms of the residence time t_e in the chamber. The values of the residence time in order to effect a certain completeness of reaction will not only be a function of the chemistry of the propellants but also of the hydrodynamic factors in injection which influence combustion rate, i.e. atomization, mixing, evaporation, etc.

The residence time is related simply to the rocket motor geometry and performance of the propellants as follows, restricted to the previous assumption of uniform flow:

$$t_c = \frac{m_c}{\dot{m}} \tag{4-1}$$

where m_c is the mass of the propellant in the combustion chamber at the steady state and $\dot{m}$ is the mass rate of flow.

Further,

$$m_c = \frac{p_c V_c \overline{\mathfrak{M}_c}}{R \overline{T}_c} \tag{4-2}$$

$$\frac{1}{\dot{m}} = \frac{c^*}{p_c A_t} \quad \text{(see Eq. 3-5)} \tag{4-3}$$

where $\overline{\mathfrak{M}_c}$ and $\overline{T}_c$ denote the molecular weight and temperature in the chamber averaged over the entire chamber contents. Eq. 4-1, 4-2, and 4-3 result in

$$t_c = \left(\frac{\overline{\mathfrak{M}_c}}{R \overline{T}_c} \right) \left(\frac{V_c}{A_t} \right) c^* \tag{4-4}$$

The quantity V_c / A_t has been termed L^* [1] and is directly proportional to residence time. Eq. 4-4 is sometimes further simplified [1] by means of a relation of the type:

$$c^* = \frac{\sqrt{RT_c / \mathfrak{M}_c}}{\Gamma(\gamma)}$$

where $\Gamma(\gamma)$ is a function only of γ (see preceding article). Identification of $\overline{T}_c$ with T_c and $\overline{\mathfrak{M}_c}$ with $\mathfrak{M}_c$ cannot be made in a strict sense, however, since the unbarred quantities refer to conditions at the entrance of the nozzle rather than averages in the chamber. Since in general T_c will be greater than $\overline{T}_c$ and $\mathfrak{M}_c$ less than $\overline{\mathfrak{M}_c}$, the following inequality can be derived:

$$t_c > \frac{L^*}{c^* [\Gamma(\gamma)]^2} \tag{4-4a}$$

The use of the parameter $L^* = V_c / A_t$ has been found very useful in the engineering development of rocket engines (see XII,G).

The result of an improperly designed combustion chamber is generally a low value of the chamber temperature T_c. The equations developed in Art. 3 for the performance parameters show that such a condition always leads to a loss in exhaust velocity. The continuation of combustion during expansion will generally permit a partial recovery of the lost combustion heat, but full recovery is theoretically impossible since the liberation of heat at lower pressure levels causes a lowering of the ideal thermodynamic efficiency. The effect of combustion during flow is to cause the rate of temperature drop to be diminished thereby decreasing the apparent value of γ. (The mechanism of this effect has been discussed more completely in the latter part of Art. 2 under the heading *Isothermal expansion*.) Incomplete combustion can generally be detected in rocket motors by the appearance of a large exhaust flame resulting from afterburning.

Reassociation reactions. The occurrence of reassociation reactions, as distinct from incomplete combustion, is not directly related to chamber design but results specifically from the changes in temperature and pressure during expansion. For high temperature propellant systems where the gases are extensively dissociated into "unstable" molecules[8] and atoms, reassociation of these components at the lower temperatures during expansion leads to a heat effect, similar to that of continuing combustion, which increases the exhaust velocity. The types of reactions which occur during expansion are listed below together with the heat release at 298°K [2].

$$H_2O + CO = H_2 + CO_2 + 9.45 \text{ kcal}$$
$$\tfrac{1}{2}H_2 + OH = H_2O + 68.10 \text{ kcal}$$
$$2H = H_2 + 103.80 \text{ kcal}$$
$$2N = N_2 + 185.1 \text{ kcal}[9]$$
$$2O = O_2 + 118.20 \text{ kcal}$$
$$NO = \tfrac{1}{2}N_2 + \tfrac{1}{2}O_2 + 21.60 \text{ kcal}$$
$$CO + \tfrac{1}{2}O_2 = CO_2 + 67.61 \text{ kcal}$$
$$H_2 + \tfrac{1}{2}O_2 = H_2O + 57.80 \text{ kcal}$$

The effect on the exhaust velocity of extensive reassociation may be as high as a 10 per cent improvement (e.g. the H_2—O_2 or H_2—F_2 propellant system) but normally does not exceed 5 per cent for carbonaceous propellants. The extent to which any one of the reactions listed above occurs is determined by the reaction rate for that equation and the residence time during flow. Since these rate laws are of the form

$$\frac{dn_i}{dt} = kf(n_1, n_2, \cdots)e^{-E/RT} \qquad (4\text{-}5)$$

where $f(n_1, n_2, \cdots)$ is some function of the molecular concentrations, it is apparent that unless complete equilibrium is maintained, the major portion of the reaction occurs in the early stages of expansion where $e^{-E/RT}$ is large, the molecular concentrations are high, and residence times are long because of low flow velocities. Theoretically, it is possible to determine the extent to which each one of these reactions occurs during expansion, by means of the equations of Art. 2 if the rate law is known. Because of the tedious procedure of calculating quantitatively the exact concentration of minor components at each stage of expansion, it is

---

[8] By "unstable" molecules is meant constituents like OH, CN, CH, NH, etc., which do not normally exist at standard temperature and pressure.
[9] This value is presently in doubt, and the value 225 kcal has recently received some support.

usually convenient to employ a criterion to determine whether or not near-equilibrium is maintained with respect to any given reaction for which the rate constants are known.

Criteria for maintenance of equilibria. For high temperature systems in which the chemical species are extensively dissociated, reassociation reactions during the expansion process result in a significant modification of the expansion law. In a formal way, the expansion law including the effect of the individual rates of reaction may be expressed as follows (cf. Eq. 2-31):

$$\sum \frac{[\Delta H_t]_i}{n} \frac{dn_i}{dt} \frac{dt}{dT} \frac{dT}{T} + \frac{C_p}{n} \frac{dT}{T} = R \frac{dp}{p} \qquad (4\text{-}6)$$

where dt/dT is the residence time of the gas per unit temperature interval. A fairly accurate evaluation of dt/dT may be made by applying an equation such as Eq. 2-32 to the flow process taking into account the geometry

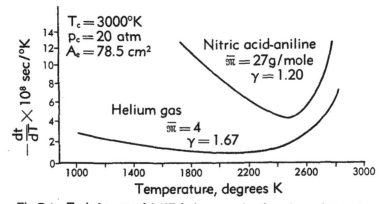

Fig. B,4. Typical curves of dt/dT during expansion through a rocket nozzle.

of the duct. For rocket motors of about 1000 pounds thrust, Altman and Penner have shown that such calculations yield residence times of the order of 10^{-8} to 10^{-7} sec/°K [8]. Fig. B,4 shows some typical curves of dt/dT during expansion through a rocket nozzle.

Because of the Arrhenius factor and the pressure effect in chemical rate expressions, reaction rates are fastest at the start of the expansion and monotonically decrease as the pressure falls. At each temperature, the residence time dt/dT can be used to determine the equilibrium values of dn_i/dt, and comparison with the kinetic values can be made to determine the approach to equilibrium. If such a comparison is made at the end state and equilibrium is followed, then it is certain that equilibrium was maintained throughout the entire expansion. Schaefer [10] and Penner [11,12] have utilized this concept to develop criteria for the maintenance of equilibria during flow through nozzles.

The treatment, essentially that of the above authors, is given here as applied to the reassociation reaction of atoms to form molecules. The method, however, can be generally applied to any reaction [12].

Consider the combination of atoms A and B with the aid of a third molecule M to form the molecule AB,[10]

$$A + B + M \underset{k_b}{\overset{k_t}{\rightleftharpoons}} AB + M \tag{4-7}$$

where k_t and k_b denote the forward and reverse (backward) reaction rates respectively. The equilibrium constant K_c for this reaction is

$$K_c = \frac{(AB)}{(A)(B)} = \frac{k_t}{k_b} \tag{4-8}$$

where the quantities in ( ) designate the concentrations of the species in moles per unit volume. Thermodynamic equilibrium during flow for this reaction can be maintained if

$$\frac{d \ln K_c}{dt} = \frac{d \ln (AB)}{dt} - \frac{d \ln (A)}{dt} - \frac{d \ln (B)}{dt} \tag{4-9}$$

The quantities $d \ln (\ )/dt$ refer to the total change of the concentrations of the species with time. Because the process is applied to an adiabatic expansion, this change results from two factors: (1) the temperature and pressure change cause a concentration change (change of state), and (2) chemical reaction causes a concentration change. The total change will therefore be the sum of these two changes.

Let $[d \ln (\ )/dt]_s$ be the fractional rate of concentration change resulting from a change in state and $[d \ln (\ )/dt]_{ch}$ that resulting from the chemical reaction. Then

$$\frac{d \ln (\ )}{dt} = \left[\frac{d \ln (\ )}{dt} \right]_s + \left[\frac{d \ln (\ )}{dt} \right]_{ch} \tag{4-10}$$

Assuming that a suitable average γ can be chosen for the expansion process, the concentration-temperature relation can be obtained directly from Eq. 2-4 noting that V is the reciprocal of the concentration. Differentiation of the logarithm of the concentration with respect to time can be accomplished by means of the temperature derivative as follows:

$$\ln (x) = \ln \text{const} + \frac{1}{\gamma - 1} \ln T \tag{4-11}$$

$$\left[\frac{d \ln (x)}{dt} \right]_s = \frac{d \ln (x)}{dT} \frac{dT}{dt} = \frac{1}{(\gamma - 1)T} \frac{dT}{dt} \tag{4-12}$$

[10] The combination of free atoms to molecules generally requires the presence of a third body to carry away the heat of reaction generated.

where x is any species in the gas. The term dT/dt is the rate of temperature change during flow, typical curves of which are given in Fig. B,4.

The concentration change resulting from chemical reaction alone is given by the rate law

$$\left[\frac{d(AB)}{dt}\right]_{ch} = -\left[\frac{d(A)}{dt}\right]_{ch} = -\left[\frac{d(B)}{dt}\right]_{ch}$$

$$= k_f(A)(B)(M) - k_b(AB)(M) \tag{4-13}$$

$$= k_f(A)(B)(M)\left[1 - \frac{(AB)}{(A)(B)K_e}\right]$$

The ratio $(AB)/(A)(B)$ at any temperature T may be replaced by the equilibrium constant K'_e at the temperature T'. For equilibrium flow, $T' = T$, whereas for near-equilibrium flow, $|T' - T|$ is a small number. Substituting K'_e for $(AB)/(A)(B)$ in Eq. 4-13 there result the following equations for the logarithmic changes:

$$\left[\frac{d\ln(AB)}{dt}\right]_{ch} = \frac{k_f(M)}{K'_e}\left[1 - \frac{K'_e}{K_e}\right] \tag{4-14}$$

$$\left[\frac{d\ln(A)}{dt}\right]_{ch} = -k_f(B)(M)\left[1 - \frac{K'_e}{K_e}\right] \tag{4-14a}$$

$$\left[\frac{d\ln(B)}{dt}\right]_{ch} = -k_f(A)(M)\left[1 - \frac{K'_e}{K_e}\right] \tag{4-14b}$$

Now, it may be noted that for near-equilibrium flow, $K'_e \cong K_e$ and therefore

$$1 - \frac{K'_e}{K_e} = -\frac{\Delta K_e}{K_e} \cong -\frac{d\ln K_e}{dT}(T' - T) \tag{4-15}$$

Substitution of Eq. 4-13, 4-14, 4-14a, and 4-14b into Eq. 4-9 by means of Eq. 4-10 yields

$$\frac{d\ln K_e}{dt} = -\frac{1}{(\gamma' - 1)T}\frac{dT}{dt}$$

$$- k_f(M)\left[(A) + (B) + \frac{1}{K'_e}\right]\left(\frac{d\ln K_e}{dT}\right)(T' - T) \tag{4-16}$$

which can be solved for $(T' - T)$ to yield

$$(T' - T)k_f(M)\left[(A) + (B) + \frac{1}{K'_e}\right]$$

$$= \left[1 + \frac{1}{(\gamma - 1)T(d\ln K_e/dT)}\right]\left(-\frac{dT}{dt}\right) \tag{4-17}$$

Eq. 4-17 may be simplified further by noting the thermodynamic relation $(d \ln K_c)/dT = \Delta E/RT^2$. Therefore $(\gamma - 1)T \cdot (d \ln K_c)/dT = (\gamma - 1)\Delta E/RT = \Delta E/C_vT$ where $-\Delta E$ is the heat evolved when the reaction occurs at constant volume of the gas. The temperature lag now has the specific expression

$$(T' - T) = \frac{[1 + (C_vT/\Delta E)](-dT/dt)}{k_t(M)[(A) + (B) + 1/K_c']} \tag{4-18}$$

Generally speaking, values of $-\Delta E$ for most atomic reactions vary from 50 to 150 kcal per mole, and in most cases, $-\Delta E > C_vT$. For the unique point where $-\Delta E = C_vT$, $T' = T$ for any value of k_t. This results from the fact that the concentration change resulting from the change in state is just sufficient to maintain equilibrium. For the case where $-\Delta E < C_vT$, the change of state causes concentration changes greater than those demanded for equilibrium and the reaction proceeds the other way, i.e. $T' < T$. The dominant terms in Eq. 4-18 are $-dT/dt$ and k_f. For rockets of about 1000 lbs thrust, $-dT/dt$ ranges from 1 to 2×10^7 °K/sec whereas for larger rockets like the V-2, the value ranges from about 1.5 to 3×10^6 °K/sec. (dT/dt is approximately inversely proportional to $\sqrt{\text{thrust}}$.) For atomic recombinations resulting from triple collisions, $k_t \sim 10^{11}$ (moles/liter)$^{-2}$ sec^{-1} [13]. Since $1 + C_vT/\Delta E < 1$, it appears that near-equilibrium can generally be maintained for atomic reactions unless the actual concentrations (A) and (B) are very small, in which case further recombination will have only negligible effect.

The equation for temperature lag of other types of chemical reactions can easily be formulated by the methods just given. The reaction types of greatest interest and their temperature lags are summarized below [12]:

1. Deactivation of excited components

$$A^* = A \tag{4-19}$$

$$(T' - T) = \frac{(-dT/dt)}{k_t(1 + 1/K_c')} \tag{4-20}$$

2. Bimolecular double decomposition

$$A + B = C + D \tag{4-21}$$

$$(T' - T) = \frac{(-dT/dt)}{k_t\{(A) + (B) + [(C) + (D)]/K_c'\}} \tag{4-22}$$

3. Double decomposition reaction[11]

$$2AB = A_2 + B_2 \tag{4-23}$$

$$(T' - T) = \frac{(-dT/dt)}{k_t\{4(AB) + [(A_2) + (B_2)]/K_c'\}} \tag{4-24}$$

[11] Eq. 4-24 contains the factor 4 in the denominator which was omitted in the original reference.

B,5. Two-Phase Flow.

SYSTEMS OF APPLICATION. The expansion process can become quite complicated if a condensed phase is present during all or part of the expansion. Such cases arise when metallic elements or salts are included in fuels or rocket propellants, in the detonation of TNT, and in the expansion of initially saturated vapors. The treatment in this section is confined to a semiquantitative discussion of the effects in open nozzles. This is considered in two parts, first for a condensed phase present in fixed amount during the entire expansion process, and second for condensation during expansion, when the amount of the condensed phase present is a function of the pressure, temperature, and rate of condensation.

GENERAL THEORY FOR PERMANENT CONDENSED PHASES. The problem of determining the course of the expansion process when a condensed phase is present may be divided somewhat arbitrarily into two parts: the hydrodynamic problem, dealing primarily with kinetic energy distribution between solid and gas, and the thermodynamic problem, dealing primarily with temperature distribution. Actually, the problems are interconnected and dependent on each other.

It is apparent that for particles of large size, a relatively long time will be required for the particles to attain the velocity and temperature of the gas. As the particle size becomes smaller, its behavior becomes gaslike and equilibration with the gas stream is very rapid. Although mathematically it is possible to represent the transition from solid (or liquid) to gas as particle size diminishes, it is known from nucleation theory that condensation requires nuclei and that the particles initially formed are of a minimum critical size dependent on the degree of supersaturation of the vapor [16,17]. In the treatment in a later part of this section, it is shown that both hydrodynamic and thermal equilibrium can be approached for particle sizes less than 10^{-4} cm in expansion processes and in regions of active combustion.

The energy equation and the "gaslike" solid. Under the assumptions that the flow process is adiabatic and the gases are perfect, the energy equation for frictionless flow is

$$C_{p_g}dT_g + C_{p_s}dT_s = V(dp_g + dp_s) \tag{5-1}$$

where the subscripts g and s refer to the gas and solid (or liquid) state respectively. The term p_s denotes the pressure exerted by the solid particles resulting from their thermal or Brownian motion. If n_s is the number of moles of solid for n_g moles of gas and if $\bar{n}_s$ is the average number of molecules per solid particle, then $n_s/\bar{n}_s$ is the number of moles of solid particles. These particles can be considered to behave like a gas of high molecular weight with the following equation of state:

$$p_b V = \left(\frac{n_a}{\bar{n}_a}\right) RT_g \tag{5-2}$$

whereas the equation of state of the gas is

$$p_g V = n_g RT_g \tag{5-3}$$

The equation of state of the particles is given in terms of T_g because the particles are assumed to be in translational equilibrium with the gas. Its internal energy, however, corresponds to another temperature T_s which approaches T_g for very high values of thermal conductivity or low time rates of temperature change. The heat capacities of n_s moles of solid are

$$C_{V_s} = C^0_{V_s} + \left(\frac{n_a}{\bar{n}_a}\right) 3R \tag{5-4}$$

$$C_{p_s} = C^0_{p_s} + \left(\frac{n_a}{\bar{n}_a}\right) 4R \tag{5-4a}$$

where the superscript ^{0} refers to the internal property of the solid (including the surface energy)[12] and the second term is the contribution resulting from the rotation and translation of the particles. For large values of $\bar{n}_a$, the two specific heats are very nearly equal, whereas for small values of $\bar{n}_a$, $C_{p_s} - C_{V_s}$ approaches R. Eq. 5-1, 5-2, and 5-3 may be combined to give

$$\left[C_{p_g} + \left(\frac{n_a}{\bar{n}_a}\right) 4R\right] \frac{dT_g}{T_g} + C^0_{p_s} \frac{dT_s}{T_s} = \left(n_g + \frac{n_a}{\bar{n}_a}\right) R \frac{dp}{p} \tag{5-5}$$

For large particles ($>10^{-7}$ cm so that $n_a/\bar{n}_a < 10^{-2}n_g$) Eq. 5-5 reduces to

$$C_{p_g} \frac{dT_g}{T_g} + C^0_{p_s} \frac{dT_s}{T_g} = n_g R \frac{dp}{p} \tag{5-6}$$

This equation cannot be integrated unless T_s is known as a function of T_g or p. Since T_s is determined by heat transfer conditions and particle size, it is obvious that there will be some lag between T_s and T_g. In the event that $T_s - T_g$ is negligible, Eq. 5-6 can be integrated directly to give ($T_s = T_g = T$):

$$\frac{T_e}{T_c} = \left(\frac{p_e}{p_c}\right)^{\frac{n_g R}{C_{p_g} + C^0_{p_s}}} \tag{5-7}$$

whereas for the other extreme where T_s remains at T_0, the initial temperature, $dT_s = 0$ and therefore

$$\frac{T_e}{T_c} = \left(\frac{p_e}{p_c}\right)^{\frac{n_g R}{C_{p_g}}} \tag{5-8}$$

represents the equation for the gas.

[12] For very small particles ($<10^{-7}$ cm), $C^0_{V_s}$ and $C^0_{p_s}$ may be expected to deviate from the values for the macroscopic solid.

The hydrodynamic problem. For larger aggregates of condensed phases, the problems may be approached from the opposite side, that of the usual laws of aerodynamics and heat transfer. These may be to a large extent separated. The force or drag acting on a spherical body with velocity relative to a gas may be represented by

$$F = C_D A \tfrac{1}{2} \rho_g (\Delta u)^2 \tag{5-9}$$

where

F = force acting on the particle
C_D = drag coefficient
ρ_g = gas density
A = frontal area of body = πr^2
Δu = relative velocity between gas and particle = $u_g - u_s$

The acceleration of the particle is determined by the force acting on it and its mass:

$$F = ma = \tfrac{4}{3}\pi \rho_s r^3 a \tag{5-10}$$

where

ρ_s = density of solid
r = radius of particle
a = acceleration of particle = du_s/dt

From Eq. 5-9 and 5-10 the particle acceleration is

$$\frac{du_s}{dt} = \frac{3\rho_g C_D}{8\rho_s r}(u_g - u_s)^2 \tag{5-11}$$

At low Reynolds numbers when the velocity lag is small ($Re < 2$), Stokes' law holds, and the acceleration is found to be

$$\frac{du_s}{dt} = \frac{9\mu}{2r^2}(u_g - u_s) \tag{5-12}$$

where μ is the gas viscosity.

In principle, Eq. 5-11 and 5-12 can be solved with the aid of the energy equation, $\tfrac{1}{2}m_g u_g^2 + \tfrac{1}{2}m_s u_s^2 = C_p(T_o - T)$ and the equation for the rate of temperature change dT/dt (see Art. 2). The equations show that small values of u_g and dT/dt favor small velocity lags, $u_g - u_s$. In nozzles, both quantities increase continually during flow and only very small particles ($<10^{-5}$ cm diameter) can be further accelerated by an appreciable amount downstream of the throat. On the other hand, particles of sizes greater than 10^{-3} cm lag considerably behind the gas velocity even upstream of the throat.

The thermal problem. The problem of temperature equilibrium between the particles and gas is considerably more complex than the hydrodynamic problem because of the additional factors that are involved. There are three major influences which must be considered here:

1. Convective heat transfer to or from the particle.
2. Heat transfer within the particle.
3. Radiative heat transfer between the particle and the surroundings.

1. Convective heat transfer. The problem of convective heat transfer from the solid particle may be treated in a manner similar to that given in the hydrodynamic treatment. This loss of heat to the gas per unit time may be represented by

$$\frac{dq}{dt} = hA(T_s - T_g) \tag{5-13}$$

where h is the average coefficient of heat transfer and A the effective surface area. Assuming instantaneous heat transfer within the particle, the rate of heat loss dq/dt can be set equal to $-C_s(dT_s/dt)$ where C_s is the heat capacity of the particle. This substitution gives

$$-\frac{dT_s}{dt} = \frac{hA}{C_s}(T_s - T_g) \tag{5-14}$$

For spherical particles, $A = 4\pi r^2$ and $C_s = \frac{4}{3}\pi r^3 \rho_s c_s$ where c_s is the specific heat per unit weight and ρ_s the density. Eq. 5-14 now assumes the form

$$-\frac{dT_s}{dt} = \frac{3h}{c_s \rho_s r}(T_s - T_g) \tag{5-15}$$

which shows that high rates of heat transfer are favored by small particles.

A simple order-of-magnitude calculation may be instructive in determining the approximate particle size necessary to maintain thermal equilibrium in nozzles. For low Reynolds numbers ($Re < 2$) the Nusselt number $2hr/k = 2$, whereas for $Re > 2$ it has a larger value. A lower limit to the heat transfer can be obtained, therefore, using the value $hr/k = 1$ in Eq. 5-15. Setting $c_s \rho_s = 1$ and using an average value for k of 3×10^{-4} cal/sec°C cm for a typical combustion gas, Eq. 5-15 reduces to

$$-\frac{dT_s}{dt} = \frac{9 \times 10^{-4}}{r^2}(T_s - T_g) \tag{5-15a}$$

For thermal equilibrium, dT_s/dt should be equal to dT_g/dt which is of the order of 10^6 to 10^8 °K/sec for conventional supersonic nozzles [8]. Taking the average value of 10^7 °K/sec, one sees that if a temperature differential no greater than 100°K is to be maintained, r must be smaller than 10^{-4} cm.

2. Heat transfer within the particle. Assuming the heat transfer process at the surface of the particle to be fast, the limitation on thermal transfer will then result from conductivity within the particle itself. An indication of the time required to establish thermal equilibrium under these conditions can be obtained from an examination of the thermal conduction equation as applied to spheres with an appropriate boundary condition.

Assume the sphere at an initial temperature T_0 immersed in a medium of temperature zero. The solution for the temperature at the center of the sphere of radius r after a time t is found to be [18]

$$\frac{T}{T_0} = 1 - \frac{r}{\sqrt{\pi \alpha t}} \sum_{n=0}^{\infty} e^{-\frac{(2n+1)^2 r^2}{4\alpha t}} \qquad (5\text{-}16)$$

Assuming $T_0 \cong 10°\text{K}$, thermal equilibrium (within $10°$) can be maintained if the time required for T/T_0 to reach a value 0.5 is approximately equal to the time required for the gas to cool $10°$. For rocket nozzles, this time may be chosen approximately as 10^{-6} sec [8]. Choosing a value of 10^{-3} cm²/sec for α, the thermal diffusivity,[13] Eq. 5-16 shows that this criterion leads to the conclusion that thermal transfer within the particle is fast for particle sizes less than 10^{-5} cm and slow for sizes greater than 10^{-4} cm. For conductive particles with $\alpha \cong 10^{-1}$ cm²/sec, the borderline for thermal transfer would lie between 10^{-3} and 10^{-4} cm, inasmuch as particle size varies as the square root of α as shown in Eq. 5-16.

3. Radiative heat transfer. Unless the gas has an absorptivity of unity, heat lost by radiation from the solid particles is largely not recovered by the system except for the case where the walls are regeneratively cooled. The limiting case of a particle radiating in a transparent gas is analyzed here.

For a particle of radius r, temperature T, and emissivity ϵ, the rate of energy loss through radiation alone is

$$-\frac{dE}{dt} = 4\pi r^2 \epsilon \sigma T^4 \qquad (5\text{-}17)$$

where σ is the Stefan-Boltzmann radiation constant. For spherical particles, the rate of temperature fall of the particle is calculated to be

$$-\frac{dT}{dt} = \frac{3\epsilon \sigma T^4}{\rho_s c_s r} \qquad (5\text{-}18)$$

Integrating between the initial temperature T_0 and T after time t yields

$$T^{-3} = T_0^{-3} + \frac{\rho \epsilon \sigma t}{\rho_s c_s r} \qquad (5\text{-}19)$$

Setting $\rho_s c_s$ and ϵ equal to unity, the equation reduces to

$$T^{-3} = T_0^{-3} + \frac{12.2 \times 10^{-12} t}{r} \qquad (5\text{-}19a)$$

Total transit times in rocket nozzles are generally of the order of 10^{-4} to 10^{-3} seconds, depending on nozzle size and composition of gases [8].

[13] The particles are generally metallic oxides or halides with low conductivity.

Assuming an initial temperature of 2500°K, the following tabulation shows the final particle temperatures attained for various particle sizes.

Table B,5a. Temperature loss for radiating particles
(initial $T_0 = 2500°K$; emissivity $= 1$; $c_s\rho_s = 1$.)

Particle radius, cm	$T°K$ $(t = 10^{-4}$ sec)	$T°K$ $(t = 10^{-3}$ sec)
10^{-2}	$\cong 2500$	2475
10^{-3}	2475	2280
10^{-4}	2280	1555
10^{-5}	1555	(698)*
10^{-6}	(698)*	—

* These temperatures are probably low because of neglect of back radiation to particle.

The data in Table B,5a show that for particle sizes greater than 10^{-2} cm, very little energy is lost through radiation under these conditions. For particle sizes less than 10^{-5} cm, however, the rate of temperature loss can exceed that of the gas during adiabatic expansion. Under these conditions, the temperature of the particle will be determined by the radiation density in the gas and by convection.

Results in limiting cases. While it is difficult to calculate the actual distribution of kinetic and thermal energy between a gas and a solid or liquid during expansion, the effect on performance of limiting cases is easily determined. The treatment for a few of these cases (1 and 4) has been given previously [19,20].

Four ideal limiting cases may be distinguished:

1. Complete kinetic and thermal equilibrium, $u_s = u_g$, $T_s = T_g$. This corresponds to considering the solid particles as large gas molecules.
2. Kinetic equilibrium, thermal insulation, $u_s = u_g$, $T_s = T_0$ (constant).
3. Thermal equilibrium, kinetic nonequilibrium, $u_s = 0$, $T_s = T_g$.
4. Complete insulation, $u_s = 0$, $T_s = T_0$. This last case corresponds to assuming very large particles.

Let

x = weight fraction of solid
m = mass of system
C_{p_g}, C_{p_s} = heat capacities of gas and solid respectively
n_g = moles of gas in the mass $(1 - x)m$

The adiabatic flow parameters for the four cases listed above may now be derived, assuming the solid particles sufficiently large so that their thermal motion can be neglected.

Case 1. $u_s = u_g$, $T_s = T_g$

The energy equation for the expansion is

$$dH = (C_{p_g} + C_{p_s})dT = Vdp \tag{5-20}$$

The equation of state, assuming the gas to be perfect, is

$$pV = n_g RT \tag{5-21}$$

Eq. 5-20 and 5-21 may be combined and integrated, assuming constant specific heat, to give the following relation between T and p:

$$\frac{T_e}{T_c} = \left(\frac{p_e}{p_c}\right)^{\frac{n_g R}{C_{p_g}+C_{p_s}}} \tag{5-22}$$

The enthalpy change between T_c and T_e is calculated to be

$$\Delta H_{T_c}^{T_e} = (C_{p_g} + C_{p_s})(T_c - T_e) = \tfrac{1}{2}mu^2 \tag{5-23}$$

from which the specific impulse may be obtained:

$$I_{sp} = \frac{u}{g} = \frac{1}{g}\sqrt{\frac{2(C_{p_g} + C_{p_s})T_c[1 - (p_e/p_c)^{n_g R/(C_{p_g}+C_{p_s})}]}{m}} \tag{5-24}$$

Case 2. $u_s = u_g$, $T_s = T_c$

The equations for this case are similar to the ones just obtained except that $C_{p_s} = 0$.

Case 3. $u_s = 0$, $T_s = T_g$

Eq. 5-22 for the T, p relation is unaltered, but since all of the enthalpy change of both gas and solid goes into kinetic energy of the gas, Eq. 5-24 is replaced by

$$I_{sp} = \frac{(1 - x)u_g}{g} = \frac{(1 - x)}{g}\sqrt{\frac{2(C_{p_g} + C_{p_s})(T_c - T_e)}{(1 - x)m}}$$

$$= \frac{1}{g}\sqrt{\frac{2(1 - x)(C_{p_g} + C_{p_s})(T_c - T_e)}{m}} \tag{5-25}$$

This particular case is approached in those systems where the particles leave the nozzle at low velocities or remain in the combustion chamber.

Case 4. $u_s = 0$, $T_s = T_c$

Eq. 5-22 and 5-25 can be employed setting $C_{p_s} = 0$.

Wimpress [20] has obtained corresponding results for cases 1 and 4, and in addition has given formulas for the discharge coefficients in these systems.

It should be noted that any temperature decrease in the solid resulting from radiation does not contribute to the momentum, and such energy, therefore, is not to be included in the equations of the previous cases.

In any real system, the solid particles will emerge from the nozzle at some fraction of the gas velocity and at some temperature between T_c and T_e. Calculations for the specific impulse for such cases can be made by means of the relations:

$$\tfrac{1}{2}m[(1 - x)u_g^2 + xu_s^2] = C_{p_g}(T_c - T_{e_g}) + C_{p_s}(T_c - T_{e_s}) \quad (5\text{-}26)$$

and

$$I_{sp} = \frac{1}{g}[(1 - x)u_g + xu_s] \quad (5\text{-}27)$$

if the ratios u_s/u_g and T_{e_s}/T_{e_g} are known.

An assumed set of conditions may be chosen to illustrate the relative importance of the effects of nonequilibrium. Let:

$T_c = 3000°\text{K}$

$p_e/p_c = 20.4$

$\quad x = 0.20$ (weight fraction of solid)

$\quad m = 1$ gram (total mass)

$\mathfrak{M}_g = $ molecular weight of gas $= 20$ g/mole

$\quad n_g = 0.80/20 = 0.04$ mole

$\quad C_{p_g} = (0.04)(15) = 0.60$ cal/deg. The assumed value of 15 cal/deg/ mole for the molar heat capacity of the gas is an apparent value (cf. Art. 2).

$\quad C_{p_s} = 0.2 \times 0.5 = 0.1$ cal/deg

The following tabulation shows the I_{sp} values for both complete thermal equilibrium and lack of thermal equilibrium for this system at various assumed particle velocities, expressed as fractions of the gas velocity.

Table B,5b.

u_s/u_g	$I_{sp}(T_s = T_g)$, sec	$I_{sp}(T_s = T_c)$, sec
0	206*	203
0.25	217	214
0.50	224	221
1.00	230	227
no solid	254	254

* Two limits are approached. For a very small velocity, where the solid moves out of chamber during time of gas efflux, $I_{sp} = 206$. If $u_s = 0$, $I_{sp} = 203$ since $T_s = T_c$ throughout flow.

From a practical point of view, the table shows that the question of thermal equilibrium is of far lesser importance than is particle velocity. For the assumed set of conditions, the extremes of possible particle temperatures result in at most a $1\tfrac{1}{2}$ per cent change in I_{sp} while the effect of velocity exceeds 10 per cent. It is interesting to note that were no solid present, the last row in the table shows that the calculated I_{sp} for the gas

⟨ 59 ⟩

at an initial temperature of 3000°K would be 254 seconds. This value may be compared with 230 seconds for the case where the gas contains a solid ($x = 0.20$) in thermal and kinetic equilibrium to show the penalty that must be paid by the appearance of a condensed phase. The introduction of metallic fuels into the combustion mixture of rocket engines to increase performance is based on the fact that the effect of increasing the initial chamber temperature will more than offset the effect of the solid particles. The net result is that a significant improvement can be obtained only when the original gas temperature is low.

FLOW WITH PHASE CHANGE.

Effects of condensation during flow. The problem of two-phase flow and of condensation during adiabatic expansion has long been known to engineers of steam turbines [6] and also to aerodynamicists in wind tunnel operations (III,F and VIII,E,3,4). More recently, rocket engineers have become interested in these phenomena as a result of the consideration of the use of metallic constituents in the propellants which generally form oxides, hydroxides, or halides in the combustion chamber. Since the phase change from gas to liquid or solid is accompanied by the release of heat and a decrease in the number of moles of gas, it is obvious that flow conditions are altered considerably when condensation occurs.

Assuming thermal and kinetic equilibrium for the condensed phase, the energy equation characterizing such flow is (effects of chemical dissociation are neglected)

$$dH = C_{p_g}dT + C_{p_l}dT + \Delta H_v dn_g = Vdp \qquad (5\text{-}28)$$

where the subscripts g and l refer to gas and liquid (or solid), respectively, and ΔH_v is the molar heat of vaporization. By applying the gas law $pV = n_g RT$ it can readily be shown that Eq. 5-28 is transformed to

$$\left[C_{p_g} + C_{p_l} + \Delta H_v \left(\frac{dn_g}{dt} \right) \frac{dt}{dT} \right] \frac{dT}{T} = n_g R \frac{dp}{p} \qquad (5\text{-}29)$$

This equation can be solved only if the quantity dn_g/dt, giving the rate of condensation or nucleation, is known as a function of the parameters of state, T and p. A considerable amount of work has been done on rates of nucleation (Sec. E) [16,17,21,22,23], but the expressions obtained for nuclei formation are too difficult for simple solution of Eq. 5-29. Qualitatively, the theory shows that the rate of condensation increases with the ratio of the actual vapor pressure to the saturation vapor pressure. During a flow process, therefore, the rate of condensation will always lag behind that value required to maintain equilibrium, and in some supersonic nozzles, may even be completely inhibited [6].

The relation between T and p during adiabatic flow will be markedly affected by the occurrence of condensation which results in a smaller value of γ, or larger value of C_p as can be seen from Eq. 5-29. From a practical point of view in rocket applications, however, there is a fortunate cancellation of effects which results in very similar values for the exhaust velocity regardless of whether or not condensation occurs. Whereas a larger value of C_p tends to increase the exhaust velocity, condensation decreases the number of moles of gas thereby lowering the expansion work. The net effect is usually to increase performance slightly when condensation occurs. As an example, if pure KCl vapor at 2300°K and at its equilibrium vapor pressure of 20 atm expands isentropically to one atm, the final temperatures are 1675°K, for condensation occurring at equilibrium, and 1180°K if the gas does not condense. However, the enthalpy changes per mole of KCl are 10,690 and 10,040 calories respectively.

Equilibrium expansion of a pure condensible vapor. The determination of the flow parameters for a pure condensible vapor under equilibrium conditions can be calculated rather simply from the vapor pressure and entropy equations. It may be instructive to examine this case because it sets limiting values to the temperature, pressure, and velocity which can be obtained for flow with condensation. At each pressure level, both the temperature and velocity will be higher than for any other nonequilibrium process.

Consider the expansion of a pure condensible vapor from the initial temperature and pressure T_0 and p_0. At some point in the expansion, say T_1 and p_1, the gas may reach its saturation pressure, and beyond that point the temperature and pressure relation will be given simply by the vapor pressure equation. The thermodynamic condition that the vapor condense during expansion is simply that the term $d(\ln p)/d(\ln T)$ for the vapor-liquid equilibrium mixture be greater than for the expansion of the pure gas. It can readily be shown that this inequality leads to the condition that $n_g \Delta H_v > C_p T$ where ΔH_v is the enthalpy of vaporization and C_p the total heat capacity.[14] For every vapor, therefore, there is a temperature $T = n_g \Delta H_v / C_p$ above which the isentropic expansion of a two-phase liquid-vapor system would lead to vaporization rather than condensation. The thermodynamic treatment in this case is analogous to the one to be given on condensation.

Because most two-phase systems in chemical equilibrium encountered in combustion chambers obey the condition $n_g \Delta H_v > C_p T$, only the condensation phenomenon is treated here. Consider condensation starting at T_1 and p_1. Since the values of T and p at any point farther down the streamline are given simply by the vapor pressure equation, the only other unknown that need be determined is the composition. This can best be done by means of the isentropic condition $\Delta S = 0$. Let X_g = the mole fraction of vapor at the temperature T, and S_g^T and S_l^T the corresponding

[14] This condition may also be applied when the condensible vapor is only a small fraction of the gas.

molar entropies of the gas and liquid. The isentropic condition then states that

$$S_g^{T_1} = X_g S_g^T + (1 - X_g) S_l^T \qquad (5\text{-}30)$$

Noting that $S_g - S_l = \Delta S_v$ the molar entropy of vaporization, Eq. 5-30 can be rewritten as

$$S_g^{T_1} = S_g^T - (1 - X_g) \Delta S_v^T \qquad (5\text{-}31)$$

and solving for the mole fraction of liquid, $1 - X_g$, there results

$$1 - X_g = \frac{S_g^T - S_g^{T_1}}{\Delta S_v^T} = \frac{R \ln (p_1/p) - \int_T^{T_1} C_{p_g} d \ln T}{\Delta S_v^T} \qquad (5\text{-}32)$$

If C_{p_g}, the gaseous molar specific heat, can be taken as constant in the interval T_1 to T, Eq. 5-32 further simplifies to the following

$$X_l = 1 - X_g = \frac{R \ln (p_1/p) - C_{p_g} \ln (T_1/T)}{\Delta H_v/T} \qquad (5\text{-}33)$$

where $\Delta S_v^T = \Delta H_v/T$ for condensation at equilibrium.

The enthalpy change per mole of mixture in this temperature interval is

$$H_1 - H = \int_T^{T_1} C_{p_g} dT + (1 - n_g) \Delta H_v^T \qquad (5\text{-}34)$$

The kinetic energy increase in this region of condensation is

$$\tfrac{1}{2} m(u^2 - u_1^2) = H_1 - H \qquad (5\text{-}35)$$

from which the specific impulse may be easily calculated.

B,6. Cited References.

1. Corner, J. *Theory of the Interior Ballistics of Guns.* Wiley, 1950.
2. Rossini, F. D., Wagman, D. D., Evans, W. H., Levine, S., and Jaffe, I. Selected values of chemical thermodynamic properties. *Natl. Bur. Standards Circ. 500*, 1952.
3. Stosick, A. J. Method used in calculating propellant properties. *Calif. Inst. Technol. Jet Propul. Lab. Progr. Rept. 1-25*, 1945.
4. Glassner, A., and Winternitz, P. F. A comparison of methods for the calculation of specific impulses in rocket motors. *Reaction Motors T.P. 1*, 1950.
5. *Jet Propulsion. Air Tech. Service Command, Guggenheim Aeronaut. Lab., Calif. Inst. Technol.*, 1946.
6. Stodola, A. *Steam and Gas Turbines*, Vol. 1. McGraw-Hill, 1927.
7. Hicks, B. L., Montgomery, D. J., and Wasserman, R. H. *J. Appl. Phys. 18*, 891 (1947).
8. Altman, D., and Penner, S. S. *J. Chem. Phys. 17*, 56 (1949).
9. Seifert, H. S., and Altman, D. A comparison of adiabatic and isothermal expansion processes in rocket nozzles. *Calif. Inst. Technol. Jet Propul. Lab. Mem. 9-15*, 1950.
10. Schaefer, K. On the thermodynamics of rocket propulsion, I. Available through Headquarters, Air Materiel Command, Wright Field, Dayton, Ohio. Feb. 1947.
11. Penner, S. S. *J. Am. Chem. Soc. 71*, 788 (1949).

12. Penner, S. S.   *J. Franklin Inst. 249*, 441 (1950).
13. Amdur, I.   *J. Am. Chem. Soc. 60*, 2347 (1935).
14. Malina, F. J.   *J. Franklin Inst. 230*, 452 (1940).
15. Zucrow, M. J.   *Jet Propulsion and Gas Turbines.* Wiley, 1948.
16. Frenkel, J.   *Kinetic Theory of Liquids.* Oxford Univ. Press, 1946.
17. Volmer, M.   *Kinetik der Phasenbildung*, 4. Bd. Bonhoeffer, Dresden and Leipzig, 1939.
18. Carslaw, H. S., and Jaeger, J. C.   *Conduction of Heat in Solids.* Oxford Univ. Press, 1947.
19. Maxwell, W. R., Dickinson, W., and Caldin, E. F.   Adiabatic expansion of a gas stream containing solid particles. *Aircraft Eng.*, 1946.
20. Wimpress, R. N.   *Internal Ballistics of Solid Fuel Rockets.* McGraw-Hill, 1950.
21. Reiss, H.   *J. Chem. Phys. 20*, 1216 (1952).
22. Fisher, J. C., Hollomon, J. H., and Turnbull, D.   *J. Appl. Phys. 19*, 775 (1948).
23. Das Gupta, N. N., and Ghosh, S. K.   *Revs. Mod. Phys. 18*, 225 (1946).

SECTION L

COMBUSTION OF LIQUID PROPELLANTS

DAVID ALTMAN
S. S. PENNER

CHAPTER 1. IGNITION PHENOMENA IN BIPROPELLANT AND MONOPROPELLANT SYSTEMS

L,1. Introduction. Stable combustion of liquid propellants in jet engines, rockets, etc., occurs at a measurable finite rate under steady state or quasi-steady state conditions. Prior to the establishment of steady state combustion, any propellant system must pass through one or more transitory stages commonly designated as the ignition process. A measure of the time required to initiate combustion is the ignition delay, the precise definition of which is a function of the empirical studies which are used for its measurement. The initial stages of steady state combustion are profoundly affected by heat and/or mass transport to the reactants, whereas these factors are of less importance during the transient pre-steady state reactions.

It is now commonly agreed that what the experimenter calls stable combustion can be described adequately only by considering both hydro-dynamic and chemical phenomena, the relative influence of the two processes varying widely from the diffusion flame to the turbulent combustion flame. A similar situation exists for the ignition phenomena. The interplay of physical rate processes such as evaporation rates, gas and liquid phase diffusion, and mixing processes on the one hand; and gas and liquid-phase chemical reaction rates with or without (surface) catalysis, on the other hand, is in general so complex that the notion of a single rate-determining step is generally not applicable, although it may still be useful in some cases. The various possible pre-steady state combustion paths may be summarized conveniently by the diagram shown in Fig. L,1 for a bipropellant system. The modifications required for a description of monopropellants are obvious. The diagram shown in Fig. L,1 is not meant to include every conceivable physical and chemical process which is of importance in defining the measured ignition delay. Thus heat transfer to and from the motor wall is not considered; nor is explicit allowance made for the importance of chamber size in determining partial pressures and hence chemical reaction and diffusion rates, or of heats of reaction in determining rates of heat transfer, diffusion, chemical reaction, and hence the rate of establishment of a steady state. It is to be recognized, of course, that the reaction scheme shown in Fig. L,1 remains equally valid for steady state combustion as for the ignition processes, the

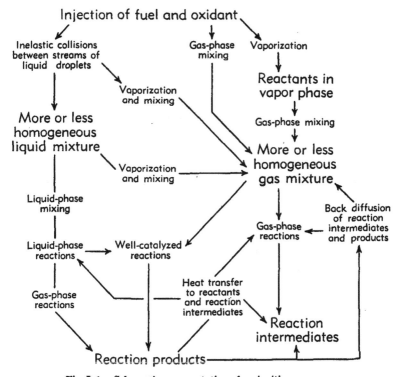

Fig. L,1. Schematic representation of preignition processes.

essential difference being that for steady state combustion the various rates of reaction are no longer changing with the time. Reference to Fig. L,1 clearly indicates the importance of both physical and chemical processes in determining ignition delay. Hence it follows that the ignition delay is in general a function not only of the propellant system used, but also of the injection procedure and the combustion chamber configuration.

At this stage of the discussion it may appear that the measurement of the ignition delay is meaningless since the ignition delay is defined through the measuring technique as well as through the physical and chemical properties of the propellant system. Although the truth of this remark can hardly be disputed, it is nevertheless known that rocket engineers have found it convenient and useful to classify propellants as (1) spontaneously-igniting or self-igniting propellants[1] and (2) non-spontaneously-igniting or non-self-igniting propellants. The mere fact that a classification of this sort is possible indicates that the injection patterns and combustion chamber configurations are sufficiently similar in many cases to permit a useful classification of propellant systems for practical purposes, or else that for many reactants relatively uncomplicated processes such as liquid-phase mixing and reactions are rate-controlling.

From the point of view of satisfactory motor performance, the use of spontaneous propellants[2] is highly desirable. Thus the temperature and pressure in the combustion chamber should increase rapidly and continuously to their equilibrium values. For nonspontaneous propellants the pre-steady state reactions may not occur at all or may occur so slowly that the propellant mixture flows through the chamber without the liberation of enough heat to initiate and sustain steady state burning. For practical purposes the ignition delay associated with this type of behavior is infinite. More frequently, undesirable motor performance is the result of excessively long ignition delays resulting in the accumulation of abnormally large concentrations of combustible gases and/or liquids in the combustion chamber before the rate of liberation of heat by the ignition processes has become sufficiently rapid to sustain steady state combustion. As the result of accumulation of abnormally large amounts of propellants prior to steady state combustion, pressure transients of disastrous magnitudes may result. This phenomenon is normally referred to as a "hard start." As a remedy for this sort of situation, the previous considerations suggest either a modification of the propellant system by the addition of a suitable catalyst or else installation of appropriate physical changes such as a better injection system to give smaller liquid droplets, more intimate liquid, and/or gas-phase mixing, etc.

The use of bipropellant systems with very short ignition delays is particularly important for service applications requiring close timing. Good ignition characteristics are also important for bipropellants used in motors which require repetitive operation.

A conservative upper limit for the transient chamber pressure during ignition, as a function of equilibrium chamber pressure, ignition delay, and motor design characteristics, can be obtained as follows. The initial mass flow rate $\dot{m}$ of propellant into the combustion chamber is given by

---

[1] The term "hypergolic," adopted from the German literature, is frequently used for "self-igniting."

[2] The use of "spontaneous propellant" for "spontaneously-igniting propellant" is a common abbreviation of the expression.

the relation

$$\dot{m} = A \sqrt{2p_i'\rho} \tag{1-1}$$

where A is the cross-sectional area of the injector, p_i' represents the injector gauge pressure, and ρ is the density of the injected fluid.[3] Actually, Eq. 1-1 overestimates the effect of $\dot{m}$ slightly since it corresponds to the equilibrium injection rate, thereby neglecting the approach to steady state flow. If it is assumed that all of the propellant remains in the chamber until ignition occurs at the time t_{ign},

$$\dot{m}t_{ign} = At_{ign} \sqrt{2p_i'\rho} \tag{1-2}$$

It is apparent that all of the assumptions made thus far tend to maximize the product $\dot{m}t_{ign}$.

During steady state combustion the required equilibrium flow rate of propellant $\dot{m}_0$ is

$$\dot{m}_0 = A \sqrt{2(p_i - p_c)\rho} = \frac{p_c f_t}{c^*} \tag{1-3}$$

where p_i and p_c represent, respectively, the absolute equilibrium values of the pressures on the injector and in the chamber; f_t is the cross-sectional area of the nozzle throat; and c^* represents the characteristic velocity. On the other hand, the maximum transient pressure p_{tr} must be less than or equal to the pressure corresponding to instantaneous combustion of all of the propellant in the chamber at the time t_{ign}.[4] Thus

$$p_t \lessgtr \frac{\dot{m}t_{ign}RT_c}{\mathfrak{M}V_c} = At_{ign} \sqrt{2p_i'\rho}\, \frac{RT_c}{\mathfrak{M}V_c} \tag{1-4}$$

where T_c is the chamber equilibrium temperature, $\mathfrak{M}$ represents an average molecular weight of the combustion products, and V_c is the chamber volume. Combining Eq. 1-3 and 1-4 leads to the result

$$\frac{p_t}{p_c} \lessgtr \frac{t_{ign}}{c^*} \frac{RT_c}{\mathfrak{M}L^*} \sqrt{\frac{p_i'}{p_i - p_c}} \tag{1-5}$$

Here the value of L^* is defined in the usual manner as the ratio of chamber volume to nozzle throat area. Eq. 1-5 can be transformed to the equivalent form

$$\frac{p_t}{p_c} \lessgtr \frac{c^*\Gamma(\gamma)}{L^*} \frac{t_{ign}}{\sqrt{(p_i - p_c)/p_i}} \tag{1-5a}$$

[3] The discussion used to estimate the ratio of transient preignition pressure to equilibrium chamber pressure will be restricted, for convenience, to monopropellants. Extension to bipropellant systems may be made without difficulty. For bipropellants, $\dot{m}$ is the sum of terms corresponding to the injection of oxidizer and fuel.

[4] Instantaneous combustion will evidently be approximated closely if the pressure rise at time t_{ign} occurs in a time interval Δt which is small compared to the equilibrium residence time in the chamber during steady state combustion.

where $\Gamma(\gamma)$ is defined as

$$\Gamma(\gamma) = (\gamma)^{\frac{1}{2}} \left(\frac{2}{\gamma + 1} \right)^{\frac{\gamma+1}{2(\gamma-1)}}$$

and γ represents the specific heat ratio.

The problem of ignition delay should not be confused with that of incomplete combustion. Thus it is well known that in some cases of steady state combustion in rocket chambers the experimentally observed performance is appreciably lower than the performance predicted theoretically from calculations assuming that thermodynamic equilibrium is reached in the combustion chamber. Although the chemical reactions occurring during the transitory ignition delay period may be in some way related to rocket performance, incomplete combustion is not considered an ignition delay problem. Nevertheless, interesting correlations have been observed by various investigators between short ignition delay and good combustion efficiency.

In conclusion it appears desirable to call attention to the fact that it is not difficult in principle to conceive an experiment in which the "true chemical ignition delay," uncomplicated by mixing phenomena, is measured. Thus it might be possible to produce essentially homogeneous mixtures of unreacted propellants at very low temperatures and then to measure reaction rates as a function of temperature for the premixed system. Studies of this type would undoubtedly be of interest to the reaction kineticist but would probably not be of immediate value to the rocket engineer interested in ignition delay measurements. At least these data would not be of practical value until the complicated interaction between physical and chemical phenomena during the ignition period is completely understood.

L,2. Experimental Methods Used for Measuring Ignition Delay in Bipropellant Systems. Experimental techniques which have been developed for differentiating between spontaneous and nonspontaneous propellants, and for estimating quantitatively the ignition delay in spontaneous propellants, fall into three general classes. These three types of tests will be designated as laboratory tests on unconfined systems, laboratory tests on semi-enclosed chambers, and motor tests, respectively.

In the laboratory tests on unconfined systems the ignition delay is usually defined by a measuring technique which emphasizes the contributions to the total ignition delay of liquid-phase mixing and reaction. Only if these liquid-phase processes occur sufficiently slowly in a rocket combustion chamber is it to be expected that laboratory estimates will agree reasonably well with observed motor performance. This type of correlation, in fact, has been observed for a number of bipropellant mixtures.

Laboratory tests on semienclosed chambers are characterized by injection systems and chamber configurations very similar to those used in actual service units. They differ from service units in that they have larger openings, and therefore maintain lower peak pressures during the ignition delay period than correspond to the equilibrium pressures which are usually encountered in combustion chamber operation.

Motor tests for determining ignition delays are carried out under conditions approximating those existing in service rocket motors. Nevertheless, scale effects on ignition reactions may be of sufficient importance to invalidate extrapolation of measurements made in small combustion chambers to performance in service units.

LABORATORY TESTS FOR MEASURING IGNITION DELAYS IN UNCONFINED SYSTEMS.

Cup tests. Among the earliest attempts made in this country at classification of propellants into spontaneous and nonspontaneous pro-

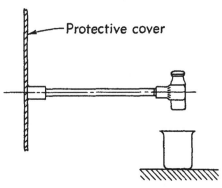

Fig. L,2a. Experimental arrangement used for cup tests. The oxidizer may be poured into the fuel or conversely.

pellant combinations, and assigning ignition delays to the spontaneously reacting mixtures, may be mentioned the simple cup-test experiments (Fig. L,2a). These tests [1,2] are performed by pouring one component of a bipropellant system into another and noting the time required for the appearance of a flame. The relative amounts of the two propellants mixed are not controlled precisely nor is there a uniform procedure for deciding which of the two components is to be poured into the other. It is usually expedient to perform several modifications of the cup-test ignition delay experiment on a given propellant mixture. In some cases the occurrence of a rapid reaction with a flame may depend critically on the manner in which the cup-test experiment is performed. More frequently, however, substantially the same results are obtained for several different procedures, thus permitting a more or less unique characterization of the propellants. In the experiments of Powell and Kaplan [1], the ignition lag in mixtures of red fuming nitric acid and aniline was found to be between 0.08 and 0.12 seconds for more than tenfold variations in the relative proportions of the added propellants. Results of this sort may be understood qualitatively in terms of an energy balance between heat liberated during reaction at the interface between the two components and the heat absorbed primarily by the propellant which is present in large excess. Both of these terms are of similar order of magnitude regardless of relative composition since the physicochemical characteristics are determined mostly by the propellants used and only to a lesser extent by the mixture ratio.

In cup test experiments the flame may be observed visually or else may be recorded by means of a high speed motion picture camera. In spite of the extreme simplicity of the cup tests, the results obtained correlate well in many cases with performance in a rocket motor under actual operating conditions. In the absence of facilities for more refined estimates of ignition delay, the cup test remains to this day one of the simplest useful methods for obtaining qualitative information on ignition delay measurements for propellant mixtures in which liquid-phase mixing and reaction contribute materially to the observed ignition delay.

Falling-drop tests. Closely related to cup tests for making preliminary estimates of ignition delay are the drop tests in which the disparity in concentration of the two components of a bipropellant system may be greatly magnified. Thus a drop of one component is allowed to fall into a cup containing an arbitrary amount of the second component.

In one type of apparatus [3,4,5,6] the falling drop interrupts the light reaching one of two photocells, thereby furnishing the reference point from which the length of the ignition delay is counted. The transitory ignition reactions are considered to be completed when the chemical reaction has become sufficiently vigorous to trigger a second photocell as the result of light emitted from the reaction mixture. The standard German drop tester was designated as the LFM Tester J. It consisted of a fuel receptacle and a calibrated orifice to release the oxidizer. The distance through which the oxidizer was allowed to fall was kept constant in order to assure similar impact conditions.

The performance of a two-photocell type of ignition delay apparatus may not be satisfactory since the second photocell is occasionally triggered by spurious radiation effects which are not related, in any obvious way, to the progress of the chemical reactions being studied. For this reason a modification of the two-photocell technique has been introduced [4,6], in which the significant physical variable determining the length of ignition delay is the time required to generate a fixed total pressure in a given combustion chamber (Fig. L,2b). Thus, Clark and Sappington define the ignition delay as the time interval between passage of the drop at the position A-A' and the time required to build up a pressure of 2 inches of water in the closed ignition chamber used for study [6]. This definition for the ignition delay represents an unfortunate choice, in that the chosen pressure is so small that the measured ignition delays appear to be abnormally short. Nevertheless the effect of various factors on ignition delay, such as composition and temperature of propellants, could still be investigated.

Another variation of the drop-test method for measuring ignition delay involves introducing one drop of oxidizer from a pipette placed a given distance above an aluminum plate containing one drop of fuel. The delay from the time of contact to ignition is measured by use of a high speed camera. Ignition is considered to occur when vigorous chemical

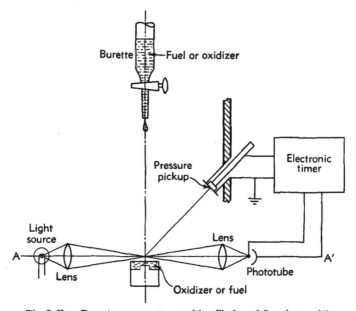

Fig. L,2b. Drop test apparatus used by Clark and Sappington [6].

reaction takes place. Whereas this method has been found to work well for the system H_2O_2-N_2H_4, CH_3OH, H_2O, satisfactory ignition does not occur for the system RFNA[5]-aniline, even though the latter system is known to be spontaneous in rocket combustion chambers. For RFNA-aniline systems, surface catalysis and liquid-phase mixing appear to be of importance in shortening ignition delay.

It is apparent that the physical variables determining the duration of the measured ignition delay in drop tests are essentially the same as the variables involved in cup tests. Considering the tenuous relation between the quantity measured in cup tests or drop tests and the empirical ignition delay which is of significance in the combustion chamber, one wonders if

[5] RFNA = red fuming nitric acid.

the refinements in measuring technique involved in the drop-test apparatus as compared with the cup tests are warranted. It appears reasonable to assume that the physical reasons leading to any significant correlation between laboratory tests and motor performance are, in large measure, the result of the fact that the rate-controlling steps in bipropellant ignition of many propellant systems involve primarily liquid-phase mixing and liquid-phase reactions. If this suggestion were not valid one would certainly expect greater disparity between measured ignition delays for changes in relative concentrations or in available surface than are actually observed.

Ignition delay measurements in the twin-jet apparatus. An apparatus for measuring ignition delay, in which the ignition delay is defined as the time required for liquid-phase mixing and sufficiently vigorous liquid-phase reaction to lead to the production of a luminous flame, has been developed by German[6] and British [4] workers. This work may be considered to carry cup tests and drop tests to their logical conclusion.

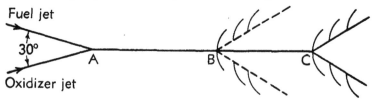

Fig. L,2c. Twin jet apparatus (schematic, cf. [4]). *C* represents the point where ignition is first observed; *B* represents the point beyond which rapid exothermic reaction occurs during the steady state. The points *B* and *C* may not coincide. The combined fuel and oxidizer jets beyond *A* are allowed to fall freely under gravitational forces.

In the twin-jet apparatus the oxidizer and fuel streams projected from two separate glass capillaries are allowed to impinge in air or nitrogen at atmospheric pressure (Fig. L,2c). The angle between the coalescing liquid propellants is chosen sufficiently small so that a single continuous stream results after collision, with relatively little carry-through. The resulting liquid mixture falls freely under gravitational acceleration until liquid phase mixing and chemical rate processes lead to vigorous exothermic reaction and vaporization of the propellant mixture. The ignition delay, as measured by the twin-jet apparatus, should then be defined as the time elapsed between impingement of the separate oxidizer and fuel streams and the *first* occurrence of rapid chemical reaction. In practice it has proved to be desirable to emphasize the point at which the ignition delay period ends by placing a strong light behind the stream of propellant mixture and measuring the length of the liquid path, which now appears as a dark silhouette against a lighter background [4].

The use of the twin-jet apparatus for the measurement of ignition

6 The German development was under way at Trauen and is quoted by Broatch [4].

delay is somewhat similar to its application for steady state combustion studies. In this case it is not the *first* occurrence of vigorous exothermic reaction which is taken as the end point; rather, a steady state measurement is made in which the position of the resulting point of vigorous exothermic reaction is dependent, to some extent, on heat and mass transfer to the liquid from the combustion products and intermediates. It is obvious that use of the twin-jet apparatus for steady state measurements will be related to ignition delay studies only if the point of vigorous exothermic reaction does not travel upstream to an appreciable extent, as the steady state is approached.

As mentioned earlier in this discussion, the twin-jet apparatus is essentially a device for measuring quantitatively the time required for the occurrence of vigorous chemical reaction as dependent on liquid-phase mixing and chemical reactions. In view of this fact it is not surprising to note that the measured ignition delay appears to increase as the angle between the two impinging propellant-component streams is decreased, thereby decreasing the rate of liquid-phase mixing. On the other hand, if the angle between the two converging liquids is made too large, considerable penetration occurs which may again lead to improper mixing and an increase in the length of the ignition delay period. For measurements on the system using 80 per cent H_2O_2-20 per cent H_2O (containing sodium pyrophosphate as stabilizer) as oxidizer and 30 per cent $N_2H_4 \cdot H_2O$-13 per cent H_2O-57 per cent CH_3OH (containing 0.6 g Cu per liter in the form of $KCu(CN)_2$ for ignition catalysis) as fuel, an angle of 30° between the impinging liquid streams was selected.

The twin-jet apparatus has also been incorporated in a large chamber in order to permit quantitative study of the effect of low pressure and temperature on ignition delay defined through the occurrence of liquid-phase mixing and reactions. These measurements are useful in assessing the importance of liquid-phase rate processes at high altitudes.

IGNITION DELAY MEASUREMENTS IN SEMICONFINED CHAMBERS. Several devices for measuring ignition delays in semienclosed chambers have been developed. In general, one would expect reasonably good correlation between this type of ignition delay test and motor performance since the various physical and chemical factors influencing the length of the ignition delay are similar.

The MIT semienclosed chamber. In an ignition delay apparatus developed at the Massachusetts Institute of Technology, a high speed motion picture camera is used to record the reaction between streams of oxidizer and fuel impinging at high velocity (approximately 150 ft/sec) in a semienclosed chamber [7]. The chamber has an opening to facilitate the photographing of the ignition reactions. The ignition delay is defined as the time elapsed between first impingement of the liquid streams and the first observation of incandescence. The end of the ignition delay period is also estimated by recording the noise accompanying the vigorous chemical reaction, with the aid of a microphone. The use of high injection velocities leads to rapid liquid-phase mixing and reactions, resulting in relatively short ignition delays.

IGNITION DELAY MEASUREMENTS IN SMALL SCALE AND SERVICE ROCKET CHAMBERS. Ignition delay measurements, published and unpublished, have been made on a great many propellant systems in a wide variety of rocket motors. A critical evaluation and compilation of the results which have been obtained is outside the scope of the present discussion. Instead, a brief review of representative apparatus for ignition delay measurements in simulated motor tests is presented.

Microrocket measurements. Ignition delay measurements in the microrocket are carried out under conditions closely approximating those occurring in service combustion chambers, the principal difference being a much larger surface-to-volume ratio and fewer injection nozzles. As with the cup-test and twin-jet measurements, results will obviously be useful if physical and chemical changes occurring in the liquid phase are rate-controlling. Otherwise, a scale effect of unknown magnitude is involved whenever an extrapolation to a larger combustion chamber is performed. This scale effect may be the result of decreased surface-to-volume ratio. Furthermore, in large combustion chambers the time required to reach the motor wall is increased, particularly for the propellants injected near the middle of the combustion chamber. As a result, increased ignition delays will be observed in large motor chambers if surface catalysis is important.

Microrockets [8] are convenient devices not only for measuring ignition delay but also for motor performance studies requiring minimal consumption of the propellants being tested. Representative microrockets have chamber lengths from $1\frac{1}{2}$ to 4 inches and nozzle throat diameters from 0.144 to 0.179 inches. A variety of injectors (Fig. L,2d) has been used. The ignition delay is defined as the time elapsed between injection into the combustion chamber and the first occurrence of a rapid increase in chamber pressure. Ignition lag measurements in the microrocket on the system H_2O_2-N_2H_4 have shown an interesting correlation between short ignition delay and high combustion efficiency.

Facilities are generally available for studying the effect of temperature, injection pressure, and initial chamber pressure on ignition lag in the microrocket.

Ignition lag measurements in transparent combustion chambers. A glass combustion chamber has been constructed in order to permit photographing of the transient chemical reactions preceding steady state combustion. The entrance of propellants, impingement, and onset of luminosity are recorded with a high speed camera. The transparent motor tests have been correlated with results obtained in metal combustion chambers in which the electrical conductivity of the liquids was used to record the time of the first impingement. The pressure rise in the combustion chamber was followed with a suitable gauge and was used to estimate the end of the ignition delay period. The injectors used for these ignition lag measurements yielded two impinging liquid streams and required between 5 and 20 cm³ of propellant per run. An apparatus of this type has been used to study the effect of temperature and of chemical additives on the measured ignition delay.

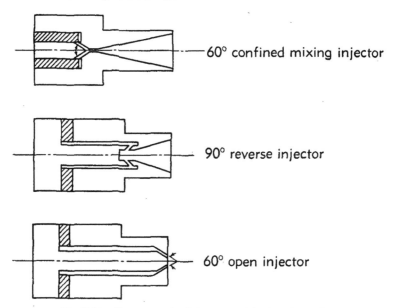

60° confined mixing injector

90° reverse injector

60° open injector

Fig. L,2d.   Representative injectors used in the microrocket.

Ignition delay measurements in service units. A wide variety of empirical information concerned with ignition delay measurements in service rockets is available. Many investigators have correlated their laboratory investigations with measurements in service units. In the more recent developments of ignition lag determinations, apparatus for systematic variations in temperature and initial pressure (in order to simulate high altitude operation) has been included. The rocket motors used for ignition lag measurements are usually small, developing thrusts in the neighborhood of 50 pounds. For this reason the problem of scale effects in designing large motors remains as an important research program to which relatively little attention has been devoted thus far.

L,3. Representative Ignition Delay Measurements on Spontaneous Bipropellant Systems. A tabulation of experimentally observed ignition delay measurements is of interest since it may aid our understanding of the transient processes preceding steady state combustion. Seven topics will be considered, viz. (1) the effect of experimental technique on observed ignition delay, (2) the effect of soluble catalysts and composition on ignition delays observed in a given apparatus, (3) the effect of initial temperature, (4) the effect of initial chamber pressure, (5) the effect of surface catalysts, (6) the relation between ignition delay and chemical structure, and (7) the effect of injection pressure on ignition delay.

Of coordinate practical importance with any of the factors enumerated above, is the relation between injection technique and ignition delay. Although the study of this problem has received considerable attention

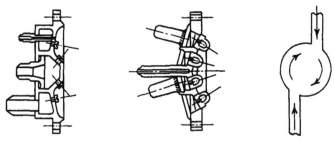

Fig. L,3a. Injectors with intersecting liquid streams which emphasize liquid-phase reactions. These injectors are useful for such systems as RFNA-AN, RFNA-N_2H_4·H_2O, H_2O_2-N_2H_4, etc. Left, intersecting stream injector; center, multiple-cone injector; right, premixing injector (the mixed fuel-oxidizer stream passes out in a direction normal to the page as a spray).

from engineers associated with motor development projects, the quantitative classification and tabulation of this type of information does not appear to be feasible. Hence a few general statements will have to suffice.

In many bipropellant systems a decrease in ignition delay, and also an increase in motor performance, may be achieved by utilizing methods for rapid and complete liquid-phase mixing. In recognition of this fact, injection techniques utilizing the type of injectors shown in Fig. L,3a and L,3b have been used in the past. Furthermore, liquid-phase premixing injectors[7] have been developed (Fig. L,3a(right)). These injectors are useful for propellant systems containing components of low volatility. On the other hand, for systems involving very volatile components such as liquid hydrogen, oxygen, fluorine, or even ammonia, short ignition delays and adequate motor performance are not critically dependent on liquid-phase mixing. In this case the injector designs drawn in Fig. L,3b(left) and L,3a(center) are useful since they give fine sprays of drop-

[7] The premixer can evidently be used for any bipropellant system provided the residence time in the injector is appreciably less than the observed ignition delay.

lets which will evaporate rapidly, thereby accelerating gas-phase reactions. Although some investigators [12] have claimed that the efficiency of the system ethyl alcohol-liquid oxygen is improved with liquid-phase mixing, this result is probably caused by the increased atomization.

COMPARISON OF OBSERVED IGNITION DELAYS FOR GIVEN PROPELLANT SYSTEMS. In order to indicate the concordance between experimental results obtained by various methods for a given propellant system, a tabulation of observed data is given in Table L,3a for the reactants

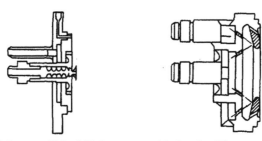

Fig. L,3b. Injectors which yield fine sprays of fuel and oxidizer, emphasizing gas-phase reactions. These injectors are useful for such systems as O_2-NH_3, H_2-O_2, etc. Left, concentric hollow cone injector; right, target type injector.

RFNA and AN. Reference to Table L,3a shows large differences in results reported by different observers although the more reliable cup and drop tests [1,5] are seen to agree well with results obtained in a semienclosed chamber and a small motor. The absolute values of the observed I.D. (ignition delay) reported in [6] are too small because a poor definition of I.D. was used (cf. Art. 2). The cup tests reported in [4] probably yielded small values for the I.D. because the reactions were catalyzed by meta

Table L,3a. Ignition delay determined by various methods for the RFNA (red fuming nitric acid)-AN (aniline) system at room temperature and an initial pressure of 1 atm.

Observed I.D., sec	Reference	Type of apparatus	Remarks
0.08 to 0.12	[1]	Cup test	The observed values of the I.D. cover a one-hundredfold variation in concentration
0.09	[5]	Drop test	
0.003 to 0.008	[6]	Drop test	The observed I.D. was roughly 2 msec larger when AN was dropped into excess RFNA, than when RFNA was dropped into excess AN
0.06	—	Semienclosed chamber and small scale motor tests	
0.004 to 0.008	[2]	Cup test	Ignition may have been catalyzed by metal salts

salts dissolved in the nitric acid. Comparing all the values obtained with the different sets of apparatus, it appears that I.D. values in the range 0.06 to 0.12 probably are most typical for this system. Some scale factor is expected when extrapolating to large service units since surface catalysis affects the initiation of the ignition reaction and surface-to-volume ratios decrease with size in geometrically similar chambers.

EFFECT OF COMPOSITION ON IGNITION DELAY. One of the important bipropellant systems for which studies of suitable ignition catalysts have been carried out is the system H_2O_2 (containing NH_4NO_3)-$N_2H_4 \cdot H_2O$. The ignition delay for this system was found to increase as the amount of ammonium nitrate added to H_2O_2 was increased [7]. Ammonium nitrate is not generally considered as a combustion catalyst, and since it does not give up oxygen as readily as does H_2O_2 its action is mainly that of a diluent.

The system H_2O_2-$N_2H_4 \cdot H_2O$ was used by the Germans in rocket air-craft and as a spontaneous combination for ignition in rocket torpedoes, with decalin employed as the fuel in the latter application for steady state combustion. In carrying out these developments the Germans found that a small amount of copper or iron catalyst, added to hydrazine hydrate in the form of complex cyanides, materially decreased the observed ignition delay. The German work has been reviewed by Darling and Kavanagh [7], who have also supplemented the available ignition delay data by carrying out measurements in semienclosed ignition chambers using photographic records to follow the progress of ignition. Representative results taken from this work are reproduced in Table L,3b [7].

Table L,3b. The effect of copper and iron catalysts on the ignition delay of 100 per cent $N_2H_4 \cdot H_2O$ reacting with 90 per cent H_2O_2 at room temperature [7].

Catalyst added	Concentration of Cu or Fe in g/l	Ignition delay, sec
$K_2Cu(CN)_4$	0.01	no ignition
	0.1	0.090
	1.0	0.021
$Na_2Fe(CN)_5NO$	0.19	0.087
	0.47	0.067
	0.93	0.036
	1.87	0.022

Although the absolute values of the I.D. listed in Table L,3b are certainly dependent on the type of ignition delay apparatus in which measurements are made, experience has shown that laboratory tests are generally useful in evaluating ignition catalysts for given bipropellant systems. Thus the use of furfuryl alcohol-aniline mixtures with RFNA and WFNA (white fuming nitric acid) is the result of laboratory measurements on the I.D. [2]. Similar remarks apply to the use of ferric chloride, vanadium pentoxide, and ammonium metavanadate used as ignition catalysts in WFNA reacting with aniline which may contain copper formate as ignition catalyst [2]. These experimental results suggest that oxidation catalysts appear to have an accelerating effect on the ignition process of these systems.

The chemical rate processes occurring either during ignition or during steady state combustion are not understood well enough to provide a complete explanation for the role of catalysts during ignition. Accordingly the development of ignition catalysts must be viewed as an art rather than as a science.

EFFECT OF TEMPERATURE ON IGNITION DELAY. Reference to the general scheme for describing ignition reactions given in Fig. L,1 suggests immediately that an increase in temperature should decrease the ignition delay since the physical and chemical factors which might conceivably be rate-controlling are all known to change as the temperature is raised. This supposition has been amply verified by studies on various bipropellant systems.

In drop-test experiments, Gunn [5] found a marked increase in I.D. for the system RFNA-AN as the temperature was decreased below about 15°C. Whereas the observed I.D. was more or less constant between 15 and 50°C at 0.09 sec, it increased to nearly 0.20 sec as the temperature was lowered to 0°C. Qualitatively similar results were obtained for the same system in drop-test experiments [6].

For a given composition of ignition catalysts used in hydrazine hydrate, Darling has observed a nearly tenfold increase in I.D. with H_2O_2 containing various amounts of NH_4NO_3, as the temperature was decreased from 25 to −35°C. These measurements were performed in semienclosed chambers. A much less pronounced dependence of I.D. on temperature has been reported for other propellant systems.

In tests performed in a 50-lb motor, an increase in ignition delay from about 0.015 to 0.040 sec has been observed as the temperature was decreased from −10 to −30°C for the system RFNA reacting with 2 parts by volume of FA plus 1 part by volume of AN. The ignition delay was observed to be constant between −10 and +10°C. Broatch [4] observed a linear decrease of I.D. in the twin-jet apparatus as the temperature was increased from 15 to 25°C for reaction between 80 per cent H_2O_2 (containing sodium pyrophosphate as stabilizer) and a fuel mixture consisting of 30 per cent $N_2H_4 \cdot H_2O$, 13 per cent H_2O, 57 per cent CH_3OH (containing 0.6 g of Cu per l in the form of $KCu(CN)_2$).

It is not possible to predict the quantitative effect of temperature on the I.D. at the present time. The data appear to indicate that not only do the rate constants influence ignition, but also physical properties such as viscosity and surface tension which affect the rate of mixing.

EFFECT OF INITIAL PRESSURE ON IGNITION DELAY. In order to simulate ignition at high altitudes, extensive experimental work has been done to determine the effect of low initial pressures, with or without simultaneous cooling, on I.D. It has been found, using the twin-jet apparatus, that spontaneous systems may exhibit sufficiently large increases in I.D. to become nonspontaneous at low pressures.

In a semienclosed chamber it has generally been found that the I.D. for systems such as RFNA-80 per cent AN plus 20 per cent FA increased as the pressure was reduced. It appears reasonable to suppose that the observed increase in I.D. for bipropellant combinations for which liquid-phase reactions are of importance is, in some measure, the result of increased droplet dispersion and vaporization at reduced pressures. Furthermore, if ignition requires the reaction of partially reacted vapors, it is clear that low pressures will inhibit this gas-phase reaction and hence increase the ignition delay.

EFFECT OF SURFACE CATALYSTS ON IGNITION DELAY. No extensive studies of surface catalysis on the I.D. of spontaneous bipropellant systems appear to have been carried out. However, this type of catalysis has been found to be effective for monopropellants such as H_2O_2 and N_2H_4 (cf. Art. 5). Accordingly one might expect that surface catalysis would also be of some value in bipropellant systems, particularly in those combinations using potential monopropellants. Failure to use surface catalysis for spontaneous bipropellants may be ascribed to the fact that other methods of accelerating reactions are more convenient and adequately efficient. As mentioned earlier in this discussion, ignition of the system RFNA-AN is facilitated by the presence of a surface. Combustion efficiency in this case is also improved by using turbulence rings.[8]

Surface catalysis has important applications in the use of nonspontaneous bipropellant and monopropellant systems (cf. Art. 4 and 5).

RELATION BETWEEN IGNITION DELAY AND CHEMICAL STRUCTURE. The relation between ignition delay and chemical structure or composition is obviously complicated by the interplay between chemical and hydrodynamic factors. Nevertheless it is to be expected that for bipropellant systems in which liquid-phase reactions are of critical importance, useful correlations can be obtained [9].

Ignition delay measurements using RFNA for the oxidizer show that unsaturated groups, such as the allyl group, enhance ignition. Saturated hydrocarbons are normally nonspontaneous. On the other hand, the aromatic nucleus has an activating effect when connected to a nitrogen or sulfur atom. Ignition is also facilitated by replaceable hydrogen atoms connected to the nitrogen or sulfur atom. These conclusions are based on cup tests, drop tests, and motor performance tests on a great variety

[8] The Enzian injector for the RFNA-N_2H_4 system makes efficient use of surface catalysis.

of organic compounds, of which mercaptans and unsaturated amines are considered to be the most promising spontaneous fuels with RFNA. The unsaturated amines were found to tolerate extensive dilution with petroleum hydrocarbons and still remain spontaneous. Preliminary screening of fuels which ignite spontaneously with RFNA was carried out by determining how far a self-igniting material may be diluted with a combustible but nonspontaneous diluent before satisfactory ignition fails to occur in drop tests. Investigations of this type are closely related to the material discussed under the heading EFFECT OF COMPOSITION ON IGNITION DELAY.

EFFECT OF INJECTION PRESSURE ON IGNITION DELAY. Relatively little quantitative work has been done on the effect of hydrodynamic factors on the ignition delay. From the earlier discussion on the mechanism of ignition delay, one might expect that high injection velocities would decrease the ignition delay because of better atomization, mixing, and frictional heating of the streams. This effect has been observed over a limited pressure range for several systems employing RFNA as oxidizer and also for the H_2O_2-N_2H_4 system in closed chambers at very high injection pressures [9]. A quantitative explanation of these effects has not been given but it appears that more experiments of this type should be useful in clarifying the influence of hydrodynamic factors on the ignition process.

L,4. Ignition of Representative Nonspontaneous Bipropellants. The discussion presented in Art. 3 suggests that it may be possible to obtain satisfactory ignition with nonspontaneous propellants by using soluble or surface ignition catalysts, by changing the injector design appropriately, or. by using an extraneous heat source to start reaction. If these methods fail, it is still generally possible to use a given non-spontaneous combination operationally by utilizing a small amount of a self-igniting combination in order to initiate combustion. An example of this procedure was quoted in Art. 3 in the use of $N_2H_4 \cdot H_2O$ to start the reaction between decalin and hydrogen peroxide in one type of German rocket torpedo. Ignition experience obtained to date indicates that this method can be successfully employed for nearly all nonspontaneous combinations containing nitric acid or H_2O_2 because these oxidizers form spontaneous combinations with many fuels.

An outstanding example of a nonspontaneous bipropellant system, which is attractive operationally because of wide availability and low cost of components, is the system RFNA-gasoline. Many soluble ignition catalysts have been developed for this system. Thus the addition of suitable amines or mercaptans to the gasoline and, alternately, the addition of a powerful oxidizing agent such as $KMnO_4$ to RFNA, results in spontaneous combinations. Unfortunately the concentrations of amines and mercaptans in gasoline required to obtain satisfactory ignition are fairly large (usually in excess of 10 per cent by volume). Furthermore the activity of $KMnO_4$, added to RFNA as an ignition catalyst, decreases rapidly with time as the result of precipitation of MnO_2. It is expected that further developments of the art will ultimately lead to the invention of more satisfactory soluble catalysts than are available at the present time.

Many bipropellant systems using liquid oxygen as oxidizer are non-spontaneous. Ignition of the system, liquid oxygen-liquid ammonia, is brought about conveniently by using a spark plug to supply the energy necessary to initiate the exothermic reactions which are characteristic of the oxidation of ammonia. High performance of this nonspontaneous system may then be obtained by using a suitable injection technique. The bipropellant combinations, oxygen-hydrazine, RFNA-alcohol, and oxygen-alcohol, can be ignited readily by using an independent energy source such as a powder squib or pyrotechnic igniter. Ignition of the oxygen-alcohol propellant in the German V-2 was initiated in this manner.

The system, RFNA-ammonia, can be made spontaneous by dissolving a small amount of an alkali metal (e.g. Li) in the ammonia.

Ignition of the nonspontaneous combination, 80 per cent H_2O_2-kerosene, may be accomplished by spraying the H_2O_2 on an appropriately heated noncatalytic surface. Catalytic surfaces for this system fall into two classes, viz. surfaces which catalyze the decomposition of H_2O_2 alone and those which catalyze both the decomposition of H_2O_2 and the oxidation of kerosene. Platinum, palladium, and hopcalite are examples of the latter while MnO_2 pellets and silver-plated copper gauze are typical of the former.

L,5. Ignition of Monopropellants. Since all useful monopropellants are intrinsically unstable, they can always be decomposed by suitable application of heat or energy. The practical requirement of short ignition delay places definite limitations on the manner in which the requisite energy is applied. A method of ignition which is suitable for one combustion chamber geometry may not be suitable for another. Consider, for instance, the ignition of a monopropellant by catalysis on the preheated walls of the combustion chamber. Heat transfer from the wall, and reaction kinetics will determine a characteristic time of ignition. This "ignition delay time," when multiplied by the injection rate of monopropellant, will give a definite quantity of propellant in the chamber leading to pressure transients which are characteristic of the chamber volume. Consider a second motor similar to the first but of shorter length, and hence smaller volume. Since the ignition geometry is substantially the same, the quantity of propellant accumulated at ignition is similar, and therefore a transient pressure greater than that occurring in the larger motor will result. Such transient pressures have been known to rupture rocket motors at starting. Thus it must be concluded that for monopropellants as well as for bipropellants not only the chemistry of the monopropellant but also the geometry of the injector and combustion chamber will influence the ignition.

The methods used for initiating the thermal decomposition of monopropellants are similar to those employed for nonspontaneous bipropellants. For convenience they are summarized explicitly below.

Heated surface. Ignition is effected by heating the combustion chamber walls or, better, by introducing into the chamber a resistance element heated electrically to a suitable temperature. This technique of ignition is applicable to nearly all monopropellants.

Pyrotechnic ignition. Pyrotechnic ignition involves the use of a small quantity of combustible placed near the injector. The heat generated by the igniter is sufficient to initiate steady state decomposition of the monopropellant.

Spark. This method, although quite commonly used for the ignition of gas mixtures, can be employed effectively with only a limited number of monopropellants. The spark is an intense temperature source confined to a small region and is usually of limited energy output. Because of the local confinement of the spark, there is danger of drowning by large droplets. Good atomization, therefore, is usually necessary for successful application of this technique.

Bipropellant ignition. This mode of ignition is restricted to those monopropellants which form spontaneously ignitable combinations with another propellant, a small quantity of which is used to initiate steady state combustion.

Surface catalysis. Chemical catalysis on surfaces may be of two types, viz. one in which the surface functions as a true catalyst and is not consumed during ignition and the other in which it reacts chemically to produce ignition. The former type of ignition is obviously more desirable since it can be used repeatedly.

Detailed information concerning specific methods for initiating the decomposition of monopropellants will be presented in connection with the discussion of the performance of these compounds.

CHAPTER 2. MOTOR PERFORMANCE OF SELECTED MONOPROPELLANTS

L,6. General Characteristics of Monopropellants. A monopropellant is a single fluid capable of undergoing exothermic reaction to yield gaseous products. The rate of decomposition of a useful monopropellant must obviously be negligible at storage temperatures. During steady state decomposition, heat and mass transport from the reaction zone occur at a rate sufficient to maintain the decomposition processes. The operation of a monopropellant rocket motor has several distinct advantages over a bipropellant motor. These are: (1) For operation of a monopropellant, one propellant tank with a single pumping arrangement is required. The simplicity so achieved may be desirable for applications where high performance is not required, as in gas generators and assisted take-off units. (2) The injection is simplified since exact impingement of propellant streams for purposes of uniform mixing is unnecessary. (3) The operation of a monopropellant motor is less apt to vary with ambient temperatures. For bipropellants, changing ambient temperatures may cause unequal density changes in the fuel and oxidizer. For a given fluid volume injected, this density change affects the mixture ratio at which the propellant operates because rates of injection of fuel and oxidizer may change differently. As a result, one propellant tank will empty prior to the other, causing an unnecessary waste of the remaining component. (4) Finally, the use of a single propellant will simplify field operations, since only one tank needs to be filled. For prefilled rockets this consideration is of little importance but it is a significant factor for units requiring tank filling in the field.

In order that a fluid be suitable for monopropellant operation, it must satisfy certain thermodynamic and kinetic requirements. The thermodynamic requirement that the substance be capable of undergoing exothermic decomposition to produce gases has already been mentioned. The kinetic requirements are generally dictated by operational procedures. The kinetic requirements are: (1) the decomposition rate at storage temperatures must be negligibly small or else capable of inhibition by chemical or other means; (2) the decomposition should be readily initiated (some of the techniques used have been discussed in Art. 5); (3) the release of energy should occur smoothly and reproducibly so that explosion or detonation can be avoided; and (4) the rate of decomposition at the steady state should be reasonably fast so that excessively large combustion chamber volumes need not be employed. The last two kinetic factors should, in principle, be capable of analysis by a theoretical treatment similar to that used to describe the burning of solid propellants. However, our knowledge regarding high temperature kinetics is so inadequate that at best only semiquantitative correlations are feasible. Since the thermodynamics of combustion is well understood, it is evident that future advances of fundamental significance in combustion will lie in expanding our knowledge of rate processes, particularly of the interplay of hydrodynamics and chemical reactions. Important steps in this direction have already been made by several investigators [10,11,12].

L,7. Classification of Monopropellants. Depending on their chemical composition, monopropellants may be conveniently divided into three separate classes.

Class A. Members of class A contain both oxidizer and fuel constituents in the same molecule. Class A embraces the large field of conventional liquid explosives, only a small portion of which has the necessary stability and kinetic requirements for use as monopropellants. Examples of class A compounds which have been considered for use as monopropellants are nitromethane (CH_3NO_2), methyl nitrate (CH_3NO_3), nitroethane ($C_2H_5NO_2$), ethyl nitrate ($C_2H_5NO_3$), nitroethanol ($HOC_2H_4NO_2$), diethylene glycol dinitrate or DEGN ($C_2H_4ONO_2)_2O$, ethylene oxide (C_2H_4O), nitroglycerin $C_3H_5(ONO_2)_3$, and hydrogen peroxide (H_2O_2).

Class B. This class contains, in general, only the separate oxidizer or fuel constituents. The decomposition energy is derived from a chemi-

Table L,7. Physical properties and heats of formation
of selected monopropellants.*

Monopropellant	Density, g/cm³	Melting point, °C	Boiling point, °C	Standard heat of formation $\Delta H^\circ_{f\,(298)}$, kcal/mole
Class A compounds				
CH_3NO_2	1.130^{20}_{4}	-29	101	$-26.7, -27.6$ [14], -21.3 [16]
CH_3NO_3	1.203^{15}_{4}	—	65, explodes	-41 (estimated)
$C_2H_5NO_2$	1.052^{20}_{20}	-90	115	-30
$C_2H_5NO_3$	1.105^{20}_{4}	<-102	88	-44.3
$HOC_2H_4NO_2$	1.270^{15}_{4}	-80	194	—
$(C_2H_4ONO_2)_2O$	1.377^{25}_{4}	-11.3	161	-103 (estimated)
C_2H_4O	0.887^{7}_{4}	-111	10.7	-18.4
$C_3H_5(ONO_2)_3$	1.601	13.3	260, explodes	-86 (calc)
H_2O_2	1.465^{0}_{4}	-0.9	152	-44.84
Class B compounds				
N_2H_4	1.011^{15}_{4}	1.4	113.5	$+12.05$
C_2H_2	0.621^{-84}_{4}	-82	-84, sublimes	$+54.19$
C_2H_4	0.566^{-102}_{4}	-169	-104	$+12.50$
NO	1.269^{-150}_{4}	-163.6	-151.8	$+21.6$
Class C compounds				
$NH_{3(l)}$	0.817^{-79}	-77.7	-33.4	-16.07 [14]
CH_3OH	0.792^{20}_{4}	-97.8	64.7	-57.04
$C(NO_2)_4$	1.650 [41]	13	126	$+8.9$ [15]

* Unless indicated to the contrary, all data in Table L,7 have been taken from the compilation of physical properties of chemical compounds prepared by the National Bureau of Standards [13].

cally unstable arrangement of the atoms in the molecule. For this reason all of the compounds in this group have positive values for the standard heats of formation (cf. Table L,7). Examples of compounds in this group, some of which may be usable monopropellants, are hydrazine (N_2H_4), acetylene (C_2H_2), ethylene (C_2H_4), and nitric oxide (NO).

Class C. A propellant of this class is a synthetic mixture of two or more compounds of fuel and oxidizer. It is evidently necessary that the components used to prepare a class C compound are not spontaneous with one another but are capable of smooth and continuous reaction once ignition has occurred. These restrictions narrow down the apparently limitless number of combinations which might otherwise be possible. Class C compounds contain the high performers among the monopropellants. Members of this class are: $NH_3(l)$-NH_4NO_3 (Diver's solution), NH_4OH-NH_4NO_3, CH_3NO_3-CH_3OH (Myrol), H_2O_2-CH_3OH, $C(NO_2)_4$-fuel.

Densities, melting points, boiling points, and heats of formation of representative monopropellants are summarized in Table L,7.

L,8. Performance Characteristics of Monopropellants. This article deals with theoretical performance values and operational characteristics of several monopropellants. The evaluation of monopropellants is concerned with performance, thermal and shock sensitivity, rate of thermal decomposition at temperatures experienced in the regenerative cooling coils, ease of combustion, and general physical properties. Next to performance, the most important of these items are the sensitivity and the thermal decomposition rate. The apparently contradictory requirements of low thermal decomposition rates at moderate temperatures, on the one hand, and ease of combustion, on the other hand, can only be realized in monopropellants with relatively high activation energies for decomposition. A suitable compromise between sensitivity and ease of combustion is not always feasible. Thus the use of a sufficiently insensitive monopropellant usually requires rather large combustion volumes leading to values for L^* which are generally considerably larger than those used for bipropellants.

PERFORMANCE OF CLASS A COMPOUNDS. A considerable amount of work has been done on the heats of combustion and the explosive properties of class A compounds, which are widely used as explosives. A comprehensive review of these compounds, as well as of class B compounds, containing an interesting correlation between oxygen balance and physicochemical properties such as power or brisance, heat of combustion, etc., has been given by Lothrop and Handrick [18]. We shall utilize the results and methods of these authors in order to correlate composition and properties for monopropellants.

The oxygen balance is defined as $-1600[2x + (y/2) - z]/($molecular

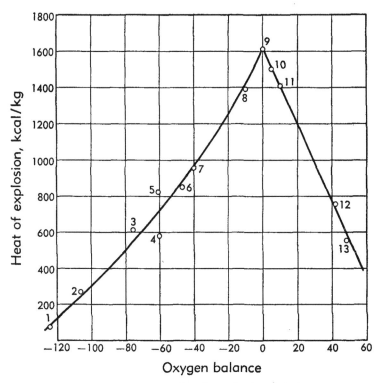

Fig. L,8a. Heat of explosion of organic nitro and nitrate compounds as a function of oxygen balance.

Legend

1. 2,3 dinitro-2,3 dimethyl butane
2. 1,3 dinitro-2,3 dimethyl propane
3. TNT
4. 2,2 dinitro propane
5. ethyl nitrate
6. tetryl
7. nitromethane

8. pentaerythritol tetranitrate
9. ethylene glycol dinitrate
10. nitroglycerin
11. manitol hexanitrate
12. hexanitro ethane
13. tetranitro methane

weight), where x represents the number of atoms of carbon, y equals the number of atoms of hydrogen, and z represents the number of oxygen atoms in the molecule. A zero balance therefore indicates that there is just sufficient oxygen to convert all the carbon and hydrogen to CO_2 and H_2O, respectively; it represents the condition for maximum heat release per gram.[9]

[9] The use of the oxygen balance ignores differences in the heats of formation of the compounds, which causes some scatter in graphs showing the desired correlation (Fig. L,8a to c). The scatter is not serious for class A compounds since the heat of combustion is much larger than the heat of formation.

The relation between heat of explosion and oxygen balance for several representative compounds of class A is shown in Fig. L,8a. The expected behavior of maximum heat release at zero oxygen balance is evident in this graph, which also shows a relatively small slope for a negative oxygen balance. This effect is far more pronounced for rocket performance parameters like I_{sp} and c^* which have a broad maximum extending toward negative values of the oxygen balance. Fig. L,8b shows the relation between explosive power as measured in the lead-block expansion test

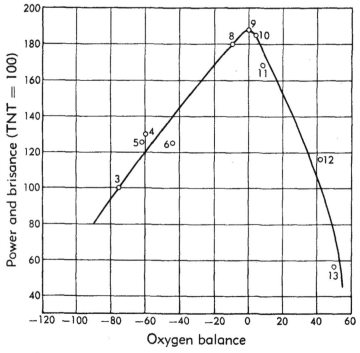

Fig. L,8b.   Power and brisance as a function of oxygen balance.
Legend: cf. Fig. L,8a.

and oxygen balance; the curve is seen to be very similar to that given in Fig. L,8a. The similarity in shape of these two curves shows that heat of explosion and explosive power are related for class A compounds. This relation is brought out in Fig. L,8c. It is of fundamental importance in evaluating class A monopropellants to recognize that high performance, which is directly related to the heat of explosion, will generally be obtained at the expense of increased explosive power.

Nitromethane.   This compound is the first member of the nitroparaffins. Thermodynamic calculations show that at the adiabatic flame temperature, the equilibrium composition is very closely represented by the following equation:

$$CH_3NO_2 = 0.2CO_2 + 0.8CO + 0.8H_2O + 0.7H_2 + 0.5N_2 \quad (8\text{-}1)$$

Using a heat of formation, $\Delta H^0_{f(298)}$ of -27.6 kcal/mole for CH_3NO_2 gives a flame temperature of 2450°K and an I_{sp} of 218 seconds at 300 psia chamber pressure exhausting to one atmosphere. The recently reported $\Delta H^0_{f(298)}$ value of -21.3 kcal/mole [16] raises the flame temperature and I_{sp} by about 4 and 2 per cent, respectively. Because of the very favorable

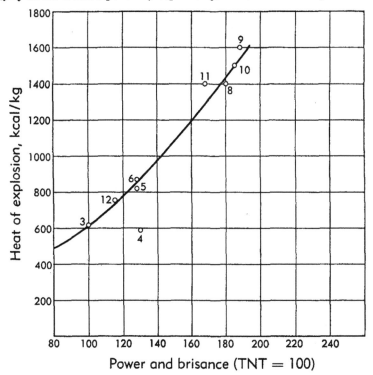

Fig. L,8c. Heat of explosion as a function of power and brisance.
Legend: cf. Fig. L,8a.

oxygen balance in this molecule, CH_3NO_2 represents one of the higher-performing monopropellants of class A.

A method which has been developed for ignition of nitromethane involves the simultaneous injection of oxygen gas and the use of a spark [17]. It is probable that the large L^* values ($\cong 500$ inches) required for continuous combustion are the result of slow gas- or liquid-phase chemical reactions rather than of slow vaporization. This supposition is supported by the fact that the use of certain combustion catalysts decreases the minimum L^* for combustion. It is generally considered, however, that such catalysts are still not sufficiently active to reduce the combustion volume to reasonably low values, say of the order of 100 inches or less. The occurrence of sporadic explosions in the regenerative cooling coils [19] is related to the use of large-L^* motors which cause a large temperature rise of the coolant. Thus the problems of increasing the combustion rate and eliminating the explosion hazard in the cooling coils are closely related for nitromethane.

Methyl nitrate. Methyl nitrate, CH_3NO_3, has a still more favorable oxygen balance than nitromethane and therefore yields higher rocket performance. Very little work has been done with this compound, however, because of its great thermal and shock sensitivity. The fact that the liquid explodes at 65°C (Table L,7) makes its use as a regenerative coolant almost impossible. Desensitizing CH_3NO_3 with CH_3OH (see Myrol in class C compounds) has been considered, but the amount of alcohol necessary for desensitization results in considerable loss in performance.

Diethylene glycol dinitrate. The oxygen balance of DEGN is about the same as that of nitromethane and hence it has very similar performance. Thermodynamic calculations [19] indicate a chamber temperature of 2500°K and an I_{sp} of 213 sec at a 20-to-1 pressure ratio. The somewhat lower performance of DEGN as compared to nitromethane results from the lower percentage of hydrogen in the molecule.

Small scale tests show that DEGN explodes more readily and at lower temperatures than does nitromethane. For this reason it is not considered possible to use this compound as a regenerative coolant.

Nitroglycerin. Nitroglycerin is only slightly overoxidized and therefore is capable of yielding high performance as a monopropellant. Its extreme thermal and shock sensitivity, however, prohibits its use as a monopropellant. Motor tests in which as much as 30 per cent nitrobenzene has been added indicate that the mixture is still too sensitive for general use. Because of its sensitivity, little development work has been carried out on this compound.

Ethylene oxide. Ethylene oxide has been considered as a monopropellant mainly because of its potential availability and low adiabatic flame temperature. The low oxygen balance (-182) indicates that the performance should be quite low. The low boiling point (10.7°C) would necessitate the use of refrigeration in moderate climates if the propellant were to be used as a liquid. Thermodynamic calculations show that the decomposition at equilibrium can be described mainly by the following reactions:

$$C_2H_4O \ (l) = CO + CH_4 + 26.6 \ kcal \qquad (8\text{-}2)$$
$$C_2H_4O \ (l) = CO + 2H_2 + C + 8.4 \ kcal \qquad (8\text{-}3)$$
$$C_2H_4O \ (l) = CO + \tfrac{1}{2}C_2H_4 + H_2 + 3.1 \ kcal \qquad (8\text{-}4)$$
$$C_2H_4O \ (l) = CO + \tfrac{1}{2}C_2H_6 + \tfrac{1}{2}H_2 + 18.9 \ kcal \qquad (8\text{-}5)$$

The relative extents to which the reactions proceed are given by the equilibria between the product species. However, because of the relatively low temperature of the reaction, it is likely that one or more of the reactions listed in Eq. 8-2 to 8-5 are inhibited as a result of slow reaction rates.

Because of the large heat release, maximum performance should result if reaction (Eq. 8-2) occurs exclusively, which would lead to a chamber temperature of 1450°K and an I_{sp} of 165 sec at a 20-to-1 pressure ratio. Laboratory experiments on the mechanism of the thermal decomposition of C_2H_4O show that free radicals are formed as intermediates and that the end products listed above are reasonable expectations [20,21]. It is apparent that the lower heat releases corresponding to reactions (Eq. 8-3, 8-4, and 8-5) will reduce performance. Because of the low performance and low adiabatic flame temperature it appears that the main field of application for this compound might be as a gas generant rather than as a rocket propellant.

Hydrogen peroxide.　With suitable catalysts it is possible to decompose H_2O_2 according to the equation:

$$H_2O_2 \text{ (l)} = H_2O \text{ (g)} + \tfrac{1}{2}O_2 + 12.96 \text{ kcal} \qquad (8\text{-}6)$$

Thermochemical calculations [22] on 100 per cent H_2O_2 show the chamber temperature to be 1250°K and the specific impulse 146 sec at a 20-to-1 pressure ratio. Because of this low performance, H_2O_2 cannot compete with the other members of the class A monopropellants, but it is interesting to note that as a result of the simplicity of operation of H_2O_2 motors and because of the low chamber temperature, H_2O_2 was actually used in military weapons by the Germans in World War II. Examples of systems utilizing H_2O_2 catalyzed by calcium permanganate are (1) the Focke-Wolf ATO Fw56 which developed 650 lb thrust for 30 seconds, (2) pilot-controlled units for the Henkel He 112 and He 126 rocket motors, and (3) the Messerschmitt Me 163-A rocket airplane with a thrust of 1650 lb [22]. More recently a group of American engineers has developed, as a hobby [23], a small monopropellant rocket of 5 lb thrust using H_2O_2. In spite of its low theoretical I_{sp}, useful rocket application can be made of this monopropellant because of its high efficiency (as measured by c^*) and its high liquid density (1.39 g/cm³ for commercial 90 per cent H_2O_2 at 20°C). Considerable application has also been made of the gas-generating properties of H_2O_2 (as for instance in the German V-2) because of ease of ignition and low temperature of the product gases.

Hydrogen peroxide is normally manufactured as an aqueous solution. Prior to World War II, concentrations no higher than 30 to 35 per cent were generally available. As a result of interest in H_2O_2 as a rocket propellant, however, commercially available solutions of H_2O_2 containing up to 99 per cent are now produced by means of vacuum distillation of the more dilute solutions. A complete review of the thermodynamic properties of aqueous solutions of H_2O_2 has been given by Kavanagh [24] and Williams, et al. [25]. For high performance it is desirable to employ as high a concentration of H_2O_2 as possible, whereas in certain gas-generating devices it may be advantageous to add water in order to lower the decomposition temperature. Calculations show that an 87 per cent solution will yield a temperature of 930°K (320°K decrease for 13 per cent H_2O) and an I_{sp} of only 126 sec (20 sec decrease). The use of aqueous solutions of H_2O_2 may be desirable because of their low freezing points.

A considerable amount of work has been done on the kinetics of the catalytic decomposition of dilute H_2O_2 solutions. It is known that permanganate solutions are effective catalysts for the decomposition of H_2O_2 and, in fact, calcium permanganate has been used very successfully for the ignition of peroxide motors [19,22]. Once the decomposition has been started, the reaction proceeds smoothly with a nonluminous exhaust. Calcium permanganate can be used in two ways, i.e. as a concentrated aqueous solution injected together with the H_2O_2, and also as a catalytic surface on alundum pellets which are soaked with the catalytic solution, dried, and then packed into the motor chamber. Both methods have been found satisfactory.

The disadvantages in the use of H_2O_2 either as a monopropellant or in a bipropellant combination are its thermal sensitivity, which requires vented containers for shipping and storage, and its high freezing point (cf. Table L,7). Concentrated H_2O_2 reacts with many metals, and the oxides formed catalyze the decomposition reaction. If poured on oxidized iron, concrete, dust, or clothing, the exothermic decomposition becomes so rapid that an explosion may result if the liquid is partially confined. This property constitutes a hazard and great care should be exercised in handling H_2O_2 to avoid spillage. Concentrated H_2O_2 is practically inert, however, in contact with 2S aluminum alloy and polyethylene, teflon and Kel-F plastics [26]. A detailed description of the equipment suitable for handling concentrated H_2O_2 has been given by Davis and Keefe [26]. With reasonable handling care and with the use of proper materials for containers these authors conclude that concentrated peroxide solutions can be handled on a large scale with safety.

PERFORMANCE OF CLASS B COMPOUNDS. The compounds in this class depend upon an unstable arrangement of the atoms for liberation of energy during decomposition. Rocket performance values for members of this group are generally lower than those of class A monopropellants.

Hydrazine. In closed-bomb tests [27] it has been shown that the thermal decomposition of N_2H_4 proceeds according to the two consecutive over-all reactions

$$3N_2H_4 = 4NH_3 + N_2 + 80.15 \text{ kcal} \qquad (8\text{-}7)$$
$$4NH_3 = 2N_2 + 6H_2 - 44.00 \text{ kcal} \qquad (8\text{-}8)$$

The NH_3 is only negligibly dissociated for temperatures below 400°K, and therefore the gases contain only NH_3 and N_2 according to the reaction given in Eq. 8-7. However, if the temperature is raised appreciably above 400°K, the NH_3 decomposes thermally according to Eq. 8-8. The equilibrium decomposition becomes essentially complete above 800°K as indicated by the data in Table L,8a which shows the equilibrium fraction x of NH_3 dissociated, originating from the decomposition of N_2H_4 at a pressure of 20 atmospheres.

Table L,8a. Equilibrium fraction of NH_3 dissociated as a function of temperature (total pressure 20 atm).

Temperature, °K	400	500	600	700	800
Fraction of NH_3 dissociated	0.05	0.25	0.60	0.88	0.96

If N_2H_4 decomposes to NH_3 and N_2 adiabatically according to Eq. 8-7, sufficient heat is liberated to heat the gases to 1649°K. If, however, equilibrium dissociation of NH_3 occurs according to Eq. 8-8, the temperature of the gases is only 867°K. Since the equilibrium concentration of NH_3 at 867°K is negligibly small, it follows that if N_2H_4 is decomposed adiabatically and permitted to reach equilibrium, the gases essentially contain N_2, H_2, and only a trace of NH_3. Suppose now that the decomposition occurs within a short period of time, of the order of milliseconds, which is characteristic of transient times in conventional rocket motors. It would then be reasonable to expect that only part of the NH_3 could decompose, depending upon the rate of the over-all reaction given in Eq. 8-8. Experimental results indicate [27] that the reaction of Eq. 8-8 proceeds more slowly than the reaction of Eq. 8-7. If the reaction of Eq. 8-7 proceeds completely and the reaction of Eq. 8-8 incompletely, then the over-all decomposition reaction can be represented as a function of the fraction x of NH_3 decomposed:

$$3N_2H_4 = 4(1 - x)NH_3 + (1 + 2x)N_2 + 6\ xH_2 + (80.15 - 44.00x)\text{kcal} \tag{8-9}$$

The performance of N_2H_4 as a monopropellant or a gas generant, therefore, is dependent on the amount of NH_3 decomposed. Some control over this factor can be had in practice through the design of motor chambers giving various L^* values.

Performance calculations [28], showing the variation of I_{sp}, T_c, and c^* with the arbitrary percentage of NH_3 decomposed, are given in Fig. L,8d. Because of the low temperatures involved and because of the high rate of travel through the nozzle, composition changes according to Eq. 8-8 would not be expected to occur. Hence constant-composition flow may be assumed. It is instructive to note that I_{sp} and c^* are quite constant over the range $0 < x < 0.5$ even though T_c is steadily diminishing. This result follows from the fact that the decomposition of NH_3 produces H_2, causing the average molecular weight to decrease about as rapidly as T_c decreases. For a monopropellant system, therefore, it is best to design the motor so that x falls in the range 0.3 to 0.5 in order to obtain maximum I_{sp} at the lowest T_c.

It is very difficult to manufacture anhydrous hydrazine because N_2H_4 forms a relatively stable hydrate $N_2H_4 \cdot H_2O$. A complete discussion of the thermodynamic and chemical properties of N_2H_4 can be found in the works of Audrieth, et al. [*29*].

The thermal decomposition rate of liquid N_2H_4 has been studied in closed-bomb tests [*27*]. It has been found that the reaction is surface-catalyzed in the gas and liquid phase, and that the hydrazine salts of HNO_3, HCl, and H_2SO_4 catalyze the decomposition in the liquid phase.

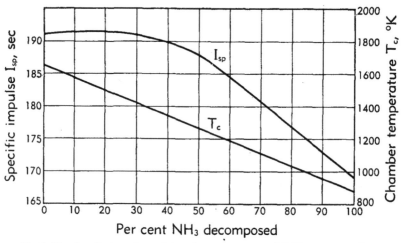

Fig. L,8d.   Specific impulse and chamber temperature of N_2H_4 monopropellant as a function of per cent NH_3 decomposed.

For uncatalyzed N_2H_4 in the constant volume system, the rate of decomposition did not vary with moderate changes in the ratio of liquid to vapor volumes, indicating that the major portion of the decomposition probably occurred in the vapor phase [*27*]. It has not been possible to detonate liquid N_2H_4 by ordinary means, and it is therefore believed that explosions experienced with N_2H_4 are probably thermal in character. This important fact makes possible the design of regeneratively cooled motors using N_2H_4, in spite of its thermal instability, since cooling passages can be designed with the maximum temperature of N_2H_4 remaining at a temperature below the value at which rapid decomposition occurs. This remark is also true for bipropellant motors using N_2H_4 as coolant, since these systems require low-L^* motors and hence lead to relatively small temperature rises of the coolant.

Acetylene, ethylene, and nitric oxide. No serious consideration has been given to the use of any of these compounds as monopropellants mainly because they are gases at normal temperatures and have several undesirable properties. Because of its large positive standard heat of formation, C_2H_2 would be expected to yield high performance. Unfortunately this compound has the serious drawback of being rather easily detonated. Both C_2H_2 and C_2H_4 would be expected to deposit carbon during decomposition. Nitric oxide has been successfully decomposed when preheated at atmospheric pressure in a flame holder, but thermodynamic calculations do not show this compound to be promising. Assuming $\Delta H_f^0(NO) = +21.6$ kcal/mole, it is found that $T_o = 2800°K$ and $I_{sp} = 190$ sec for a 20-to-1 pressure ratio employing NO gas at 300°K

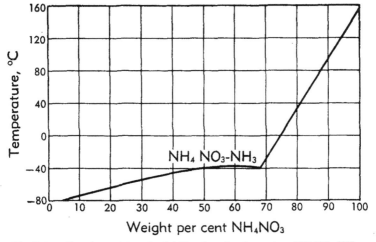

Fig. L,8e. Freezing point and solubility data for the system NH_4NO_3-NH_3.

initially. The low performance for such a high flame temperature results from the high molecular weight of the product gases.

PERFORMANCE OF CLASS C MONOPROPELLANTS.

Diver's solution. Diver's solution is a solution of NH_4NO_3 in liquid NH_3. Interest in this mixture was stimulated by the fact that the components are cheap and readily available commercially, and also because of the high solubility of NH_4NO_3 in NH_3. The freezing point or solubility curve for the system NH_3 (l)-NH_4NO_3 is shown in Fig. L,8e. Since the stoichiometric concentration occurs at 87.5 per cent NH_4NO_3, it is obvious from the solubility curve that the practical application of these solutions is restricted to the underoxidized side. The solubility at 0°C is about 75 per cent NH_4NO_3 by weight; the I_{sp} calculated at this point is approximately 175 sec at a 20-to-1 pressure ratio; the density at 20°C is about 1.15 g/cm³.

Practical application of Diver's propellant, however, has been retarded by problems of ignition and stable combustion. It is known that propellants containing NH_3 or NH_4NO_3 are difficult to ignite. Since the combustion occurs only when both components are heated to a temperature considerably above the boiling point of NH_3, it is obvious that rapid liquid-phase reaction cannot occur without catalysis. Evaporation and heating of the separate components appear to be prerequisites for both satisfactory ignition and combustion processes. Successful applications of Diver's solution depend on the development of a powerful catalyst for the combustion and a suitable arrangement of surfaces to effect a large heat transfer rate to the propellant.

Methyl nitrate and methanol (Myrol). As mentioned earlier in discussing methyl nitrate as monopropellant, the addition of CH_3OH to CH_3NO_3 has been considered in order to desensitize the methyl nitrate. The Germans did a considerable amount of work with these propellant mixtures during World War II (the term Myrol was invented by them) but the general conclusion appears to be that organic nitrates such as CH_3NO_3 cannot be desensitized sufficiently to lead to hazard-free operation, particularly in regeneratively cooled motors.

Peroxide and methanol. The purpose of adding a fuel to H_2O_2 is to bring the system closer to stoichiometric mixture ratio, thereby improving performance. Thus the addition of 21.4 per cent by weight of CH_3OH to 87 per cent H_2O_2 will raise the performance from 126 sec to 225 sec [19]. A serious drawback, however, is the great sensitivity of mixtures such as these. The presence of a fuel greatly increases the exothermic heat release during thermal decomposition, thereby making thermal explosions highly probable. The stoichiometric mixture referred to above can be detonated in a beaker by the addition of a few drops of permanganate solution which initiates the thermal decomposition of the peroxide. Because of this great sensitivity, monopropellants containing peroxide and a fuel cannot be recommended.

Tetranitromethane and fuel. Tetranitromethane, $C(NO_2)_4$, because of its high oxygen balance is a poor monopropellant. When mixed with a fuel, however, the heat of combustion is increased as is also the explosive power (cf. Fig. L,8b and c). Table L,8b shows some results obtained by

Table L,8b. Explosive power of tetranitromethane mixtures.

Detonator cap no. (German standard)	TNM	TNM:N₂O₄ 70:30	TNT	Nitroglycerin	TNM:benzene 87:13
1	12	0	0	171	413
2	44	0	208	172	404
3	65	13	218	379	404
5	86	48	268	407	404
8	71	37	332	397	445

the lead-block expansion test on the explosive power of several TNM mixtures [30], compared with TNT and nitroglycerin.

Reference to Table L,8b shows that the presence of only 13 per cent by weight of benzene causes the explosive power of TNM to surpass even that of nitroglycerin, a result which is in accord with the correlation presented in Fig. L,8b. Hence mixtures of TNM and a fuel cannot be considered to be suitable monopropellants. Pure TNM, however, is a relatively low order explosive which can be further desensitized by addition of N_2O_4.

CHAPTER 3. COMBUSTION OF SELECTED BIPROPELLANT SYSTEMS

L,9. Classification of Oxidizers and Fuels. Bipropellant combinations for use in rocket engines are derived from a relatively small group of oxidizers and a large group of fuels.

An oxidizer is a substance containing an excess of oxygen or halogen; fluorine and chlorine are the most useful halogens from an energetic point of view. The fuels generally contain carbon, hydrogen, and light metals. An element such as sulfur occupies a more or less neutral position and is considered an oxidizer in the presence of excess fuel (where it forms hydrogen or metal sulfides), but acts as a fuel in excess oxygen (where it forms SO or SO_2 during combustion).

Oxidizers which have been considered for rocket applications include O_2, O_3, H_2O_2, HNO_3, N_2O_4, NO, N_2O, $C(NO_2)_4$, F_2, ClF_3, and OF_2. Of these oxidizers, O_2, O_3, NO, N_2O, F_2, and OF_2 are gases at ambient tem-

Table L,9. Properties of representative oxidizers.

Oxidizer	Formula	Density,* g/cm³	Melting point, °C	Boiling point, °C	Standard heat of formation $\Delta H^0_{f(298)}$, kcal/mole [13]
Oxygen	O_2	1.14^{-183}	-218.7	-183.0	0
Ozone	O_3	1.71^{-111}	-251.4	-111.5	34.0
Hydrogen peroxide	H_2O_2	1.44	-0.1	150.5	-44.84
Nitric acid	HNO_3	1.50	-41.6	86.0	-41.40
Nitrogen tetroxide	N_2O_4	1.44	-11.2	21.2	2.3
Nitric oxide	NO	1.27^{-150}	-163.6	-151.8	21.6
Nitrous oxide	N_2O	1.23^{-89}	-102.4	-88.5	19.5
Tetranitromethane	$C(NO_2)_4$	1.65	13	126	8.9 [15]
Fluorine	F_2	1.55^{-187}	-217.9	-188.0	0
Chlorine trifluoride	ClF_3	1.77^{12}	-82.6	12.1	37.0
Oxygen bifluoride	OF_2	1.90^{-224}	-223.8	-144.8	5.5

* At 20°C, unless otherwise specified by the superscript.

peratures and require refrigeration or high pressures for liquefaction. The physical properties of these oxidizers are summarized in Table L,9.

The number of compounds containing carbon and hydrogen which can act as fuels is almost limitless and includes practically all of the oxygen-deficient organic chemicals. In addition, a large group of fuels containing inorganic hydrogen compounds can be employed. Thus fuels could be subdivided conveniently into organic and inorganic compounds. The organic fuels may be classified further according to the conventional groupings of organic chemistry, i.e. saturated and unsaturated hydrocarbons, alcohols, amines, ethers, nitroparaffins, aromatic compounds, etc. Important inorganic fuels include H_2, NH_3, N_2H_4, PH_3, H_2S, B_2H_6, Si_2H_6, $LiNH_2$, etc., and some of the light metals, their alloys, and solutions. The highest performing members in this group are H_2, N_2H_4, NH_3, and the hydrides and amides of the low atomic weight metals.

L,10. Combustion of Bipropellants in Rocket Engines. The physicochemical phenomena involved in the analysis of combustion processes in bipropellant motors are so involved that a quantitative discussion does not appear to be feasible at the present time. However, it may be instructive to classify the more important processes which must take place and to present simplified physical pictures for some of the complicated phenomena associated with heterogeneous combustion.

In general, bipropellant propulsion systems involve injection of liquid oxidizers and fuels. The physicochemical processes leading to the formation of the combustion products may be represented by a reaction scheme which is similar to that used in categorizing the rate processes involved in ignition (cf. Fig. L,1). The major difference between the ignition and steady state burning processes arises from the fact that heat transfer from regions of active combustion and diffusion of combustion intermediates play a more important role during steady state burning than during ignition. This difference has been emphasized previously in the discussion of measurement of ignition delay by the twin-jet apparatus (cf. Art. 2). Only a small change with time in flame position is observed during ignition for the system N_2H_4-H_2O_2. On the other hand, for the bipropellant system acid-aniline, the flame position changes markedly after ignition occurs on a surface. Hence it follows that the course of combustion during ignition and steady state burning need not be the same. This result is contrary to the idea that ignition delays are necessarily related to steady state combustion times or motor sizes required to obtain optimum performance.

During ignition of spontaneously-igniting propellants, an initial exothermic reaction generates the heat necessary for the propellant to proceed through the combustion reactions. During steady burning the flame front must be considered as an additional source of heat. In the case of a nonspontaneous system such as HNO_3-hydrocarbon, for instance, an injector which can rapidly vaporize the propellant is found to be efficient since the liquid-phase mixing can contribute only negligible heating. On the other hand, a spontaneous system such as HNO_3-aniline is found to combust more efficiently with liquid stream impinging-jet injectors. Still another situation arises with a system such as HNO_3-N_2H_4 which is spontaneously-igniting. Combustion efficiency in this system is highest when an Enzian-type injector (Fig. L,3b) is used where the propellant impinges on a surface. In this case, the initial exothermic reactions are hastened by the presence of a surface, an observation which is in agreement with the fact that the thermal decomposition of N_2H_4, which is exothermic, is surface-catalyzed [27].

In a general way, these principles concerning the path of combustion lead to some conclusions concerning the influence of chemistry on the burning process. If combustion is initiated in the liquid phase as a result of spontaneous chemical reaction, then chemical additives which alter the ignition delay may be expected to influence the over-all combustion time. Thus for HNO_3 and aniline, the presence of water in the acid increases the ignition delay and also the motor length (L^*) for steady state combustion, whereas the presence of metallic additives has an opposite effect (Table L,3a). Thus it has been observed that the presence of the corrosion products of steel (Fe, Ni, and Cr nitrates) in the acid increases the efficiency of combustion. In the case of the nonspontaneous systems such as O_2 and alcohol, however, as much as 25 per cent H_2O can be added to the alcohol without seriously altering the necessary motor dimensions. A small effect can be attributed mainly to the lower flame temperature and hence lower rate of heat transfer back to the initial combustion phase. One may further conclude from these remarks that the use of detergents in the propellants would have a more pronounced effect with systems requiring liquid-phase mixing since the ignition delay may be lowered by this means.

The theoretical calculation of combustion rates in bipropellant motors must be regarded as one of the most important unsolved problems in rocket engineering. If it were obvious that any particular process is slow enough to control the over-all reaction rate, then it would not be difficult to obtain approximate analytic expressions for the combustion rate. For example if the slow step involved diffusion of reactants, reasonable approximations could be derived from the kinetic theory of gases. If the chemical reactions between liquid propellants are important, an analysis of liquid-phase mixing and liquid-phase chemical reaction rates is indicated. If the liquid-phase reaction processes are slow compared with gas-phase reactions, evaporation rates may become rate-controlling and useful estimates of over-all reaction rates can be derived from application of formulas for droplet burning in reacting gases.

The a priori classification of rate-controlling steps for bipropellant systems is, at best, hazardous. An attempt to analyze heterogeneous combustion between gaseous oxygen and liquid hydrocarbons under various simplifying assumptions has been made by Spalding [33,34]. The conditions in bipropellant rocket motors do not, in general, correspond to the idealized treatment given by this author. Nevertheless an extension of the ideas proposed by Spalding may be useful in some cases. For example, in an ammonia-oxygen motor one would expect relatively rapid evaporation of the oxygen, leading to combustion conditions which may be approximated by the heterogeneous reaction between liquid ammonia and gaseous oxygen. For a process of this type, the notion of a "combustion surface" or a "small combustion volume" may be useful. The over-all reaction rate is then determined by the following physico-chemical processes:

1. Diffusion of gaseous oxygen and ammonia to the combustion surface which has become established close to the droplet of liquid ammonia.
2. Heat transfer to the liquid ammonia droplets and to gaseous oxygen.
3. Evaporation of liquid ammonia, which depends on the heat transfer to the ammonia and on the evaporation coefficient of the liquid.
4. Chemical reactions between gaseous ammonia and oxygen at the "combustion surface." In the treatment of Spalding, the chemical reactions are assumed to occur so rapidly that they do not contribute appreciably to the reaction time.

The diffusion of reaction products from the "combustion surface" will influence the over-all reaction rate indirectly through its effect on the distances from the "combustion surface" to the surface of the liquid and to the unreacted oxygen. The preceding picture is based on the assumption that reactions in an ammonia-oxygen motor can be represented by the heterogeneous combustion of liquid ammonia in an atmosphere of gaseous oxygen. Injection conditions also enter into the formulation of the problem indirectly insofar as they affect the distances from the "combustion surface."

Without analytic study of the results derivable from these assumptions it is not possible to assess the validity of the proposed picture. However, in a field as complicated as the combustion processes in rocket motors, it is perhaps justifiable to commence analytic treatment by using somewhat drastic simplifying assumptions. Combustion models for other types of heterogeneous combustion can be obtained by an obvious extension of the model proposed for the ammonia-oxygen system. In concluding this discussion, mention should be made also of important contributions to the problem of heterogeneous combustion which have been made recently by workers at the National Gas Turbine Establishment of Pyestock, Hants, England. For example, Barr and Mullins [35] have studied the effect of addition of combustion products to air in air-fuel combustion, Godsave has investigated the burning of single drops of fuel both theoretically [36] and experimentally [37], and Mullins has studied the ignition of fuels injected into a hot air stream [38,39]. Of particular interest is an experimental study by Topps [40] on the evaporation and combustion of falling fuel droplets which shows that the mass evaporated at any given time, divided by the product of the frictional drag on the droplet and the diameter raised to the −0.3 power, is a unique function of the temperature for a wide variety of hydrocarbon fuels.

L,11. Performance Characteristics of Several Bipropellant Combinations. In order to illustrate the behavior of representative bipropellant combinations, the performance of the following systems will be considered: RFNA-aniline; RFNA-hydrazine; RFNA-gasoline; hydrogen-fluorine. Before proceeding with the discussion of individual propellants, it may be of interest to summarize the desirable properties of propellants used in liquid-fuel rocket engines. Such a summary is presented in Table L,11.

RFNA-aniline. Theoretical performance data, assuming a 20-to-1 pressure ratio with unit atmospheric pressure, indicate a maximum I_{sp} for equilibrium flow of 226 sec for RFNA-aniline. The stoichiometric mixture ratio r (weight of oxidizer/weight of fuel) occurs at a value of 4.17 whereas the maximum value of I_{sp} (221 sec) for frozen nozzle flow occurs at a lower mixture ratio. The maximum value of I_{sp} (226 sec) for equilibrium expansion through the nozzle lies closer to the stoichiometric mixture ratio, a characteristic which is general for all systems investigated thus far.

Ignition delay measurements for this system give a value in the vicinity of 60 milliseconds in a small scale injection apparatus (cf. Table L,3a). The fact that a multiorifice-type injector gives good performance indicates that liquid-phase reactions are important initially. These reactions are probably of three types [41, pp. 610–616]: (1) neutralization of the amine group, (2) oxidation of the amine group to form nitroso- and diazo compounds which are highly colored, and (3) nitration of the benzene ring. These reactions are sufficiently exothermic to initiate steady state combustion and probably account for the observed spontaneous reaction between acid and aniline. It is known that the nitration reactions of aromatic nuclei are facilitated by the presence of NO_2^+ present in concentrated nitric acid [42,43] and that the presence of water in the acid suppresses the formation of NO_2^+. This fact is in accord with the observation that the addition of water to the acid materially increases the ignition delay and decreases the combustion efficiency.

RFNA-hydrazine. This spontaneous combination is representative of the class of noncarbonaceous propellants. The relatively high performance in this system ($I_{sp} = 246$ sec for equilibrium flow) results because of the high energy release in the reaction (N_2H_4 itself is a monopropellant) and because of the low average molecular weight of the combustion gases. The small value of r (1.56) for the stoichiometric mixture is also characteristic of most noncarbonaceous systems.

Table L,11. Desirable properties of propellants
used in liquid-fuel rocket engines.

1. Small negative or preferably positive standard heats of formation, $\Delta H^0_{f(298)}$, of the propellants are desired for high performance. However, stability requirements may limit the use of compounds with large positive values for ΔH^0_f.
2. The reaction products should have low molecular weights and large negative heats of formation. If conditions (1) and (2) are met, then the reaction products will consist of low molecular weight compounds at high temperatures.
3. The propellants should have large densities in order to minimize the dead weight of storage tanks.
4. The oxidizers and reducing agents are best handled as liquids. Hence it is desirable to obtain propellants which are normally liquid in the operating temperature range of service units (i.e. from about -40 to $+60°C$). For substances such as liquid oxygen and hydrogen, special equipment must be provided which represents added dead weight which the propulsion unit must carry and is warranted only in the case of very high energy propellant mixtures.
5. For purposes of regenerative and sweat cooling of the combustion chamber, the propellant used for cooling should have a high specific heat and a high heat of vaporization; it should be thermally stable at elevated temperatures and also noncorrosive.
6. Since it may be necessary to store the propellants for long periods of time before use, good propellants should have high storage stability, i.e. they must not decompose or change chemically during storage so that their use as a propellant is impaired.
7. Since propellants are chemicals which have to be handled by service personnel it may be desirable for some applications to use propellants of relatively low toxicity.
8. For repetitive motor operation, it is desirable to consider bipropellant mixtures which are spontaneously combustible with minimum time lag. These may be used as the main propellants or in small quantities as starting fluids.
    In addition to the items listed here, other specific considerations such as erosion of the combustion gases on the nozzle throat, deposition in the nozzle throat of solid deposits, effect of viscosity on line pressure losses, radar attenuation of the jet gases, and the economics of the propellant must be considered.

Although the ignition delay for this propellant system is quite short, the use of the multiorifice injectors employing liquid stream impingement does not result in optimum performance. However, if liquid stream impingement occurs on a surface, then the combustion efficiency is greatly improved (cf. Art. 10). The initial rapid exothermic reaction is probably the neutralization reaction which liberates sufficient heat to facilitate oxidation. When heated, hydrazine will decompose to NH_3 and the elements; and if insufficient mixing between oxidizer and fuel droplets occurs, the droplets with excess fuel will generate NH_3 which must subsequently be oxidized in the vapor stream. If the initial neutralization reaction occurs on a surface, however, then the probability of local inhomogeneities is diminished and the rate of oxidation of the N_2H_4 or NH_3 will be increased because of surface catalysis. The oxidation of NH_3, in particular, is known to be surface-catalyzed.

RFNA-gasoline. The propellant system RFNA-gasoline is of great practical utility because of the relatively large availability of the components. Gasoline is a mixture of various hydrocarbons in the range of five to twelve carbon atoms. The average composition is generally taken as C_8H_{18}. Performance calculations on this propellant combination indicate a slightly higher performance (by several seconds in I_{sp}) than for the acid-aniline system. The stoichiometric mixture ratio, $r = 5.5$, is rather high, but maximum performance occurs near $r = 4$. The propellant system is nonspontaneous and therefore requires an external ignition source [44].

If nitric acid and gasoline are mixed in a beaker, exothermic chemical reaction occurs. Depending upon the quality of the fuel, the temperature may either rise very rapidly with considerable fuming of the mixture or else the temperature rise may be practically nonexistent. This reaction is associated with nitration of the hydrocarbon molecules, which is exothermic and which proceeds more rapidly with aromatic and unsaturated hydrocarbons than with the straight-chain saturated molecules. The order of reactivity generally follows the scheme: straight-chain < branched-chain < aromatic < unsaturated compounds. The unsaturated hydrocarbons (generally olefins) undergo several types of reactions of the type [45, pp. 110–549]:

$$\underset{}{-\overset{\displaystyle H}{\underset{\displaystyle }{C}}=\overset{\displaystyle H}{\underset{\displaystyle }{C}}-} + HNO_3 \rightarrow -\overset{\displaystyle H}{\underset{\displaystyle H}{C}}-\overset{\displaystyle H}{\underset{\displaystyle ONO_2}{C}}-$$
alkyl nitrate

$$-\overset{\displaystyle H}{C}=\overset{\displaystyle H}{C}- + HNO_3 \rightarrow -\overset{\displaystyle H}{\underset{\displaystyle OH}{C}}-\overset{\displaystyle H}{\underset{\displaystyle NO_2}{C}}-$$
nitro alcohol

$$-\overset{\displaystyle H}{\underset{\displaystyle OH}{C}}-\overset{\displaystyle H}{\underset{\displaystyle NO_2}{C}}- + HNO_3 \rightarrow -\overset{\displaystyle H}{\underset{\displaystyle ONO_2}{C}}-\overset{\displaystyle H}{\underset{\displaystyle NO_2}{C}}- + H_2O$$
nitro alkyl nitrate

All of these reactions are exothermic and probably account for the large temperature rise which occurs when nitric acid is mixed with unrefined gasoline. In a study of the chemical behavior of solids formed in the *initial reactions between* WFNA and *dicyclopentadiene*, Trent and Zucrow [46] have concluded that the nitro and nitrate groups are formed. Although the nitration reactions probably constitute an initial step in combustion, their occurrence at low temperatures appears to be undesirable from an operational point of view. Thus, if a long ignition delay occurs, these reactions cause the accumulation of nitrated substances which can subsequently explode. With straight-chain saturated hydrocarbons, however, the amount of nitrated products formed during a long ignition delay is small, thereby minimizing this hazard.

Hydrogen-fluorine. The hydrogen-fluorine propellant combination is one of the highest performing systems available. Performance calculations for equilibrium expansion through the nozzle for a chamber pressure of 300 psia and an external pressure of 1 atm yield $I_{sp} = 365$ sec for $r = 3.3$, i.e. an appreciable excess of H_2 compared to the stoichiometric mixture ($r = 19$).

Although the adiabatic flame temperature at 20 atm for the H_2-F_2 system is very high (ca. 4800°K), the total equilibrium radiant heat transfer from the combustion products is not very large. The only molecule present in the combustion chamber which makes an important contribution to the equilibrium radiant heat transfer is HF. Rough estimates [47] indicate that the radiant heat transfer to the motor chamber from a H_2-F_2 motor will not exceed about 20 per cent of the total observed heat transfer.

The H_2-F_2 system is nonspontaneous, a result which has interesting chemical implications. Thus it has been shown, both from an experimental study [48] and on theoretical grounds [49], that the activation energy for the direct reaction between the two molecular species must be quite large. This work suggests that one of the functions of the igniter, such as a squib, is the production of minute concentrations of atomic hydrogen and/or fluorine. Thus the igniter may be considered to be a chain initiator, rapid reaction occurring once a small amount of the chain carriers has been produced.

The reaction kinetics of the H_2-F_2 system is relatively so simple that an approximate analysis of composition changes during nozzle flow can be carried out. The reaction scheme is

$$H + H + M \rightarrow H_2 + M$$
$$F + F + M \rightarrow F_2 + M$$
$$H + F + M \rightarrow HF + M$$
$$H + F_2 \rightarrow HF + F$$
$$H_2 + F \rightarrow HF + H$$

or for the over-all process:

$$H_2 + F_2 = 2HF$$

where M represents any third body in a collision process. This seems to be compatible with the maintenance of chemical equilibrium during expansion through a Laval nozzle [50].

L,12. Modern Trends in Combustion Research on Liquid-Fuel Rocket Engines. In recent years there has been increased activity on various phases of combustion research relating to rocket motor development. A review of some of this work is given in a recent survey paper [51] on fundamental studies of droplet burning, spray combustion, observations in motor chambers, and instability. In a series of four papers presented at the Fourth Symposium on Combustion [52, pp. 818-864], both theoretical and experimental data were presented for single fuel droplets burning in atmospheres containing oxidizer. These papers give equations for the mass burning rate as a function of the physical and thermodynamic properties of the fuel as a function of pressure up to about 20 atmospheres, and as a function of Reynolds number of the gas based on the droplet. Consideration is also given to the problem of the time required to ignite a droplet when subjected to a heated gaseous atmosphere. In several later papers [53,54,55,56], the droplet burning equations are examined in greater detail. Recognition is taken of the fact that the combustion surface is established at that point where the mass flow rates of fuel and oxidizer meet in stoichiometric proportions. This concept has also been expressed by Spalding [33; 34; 52, pp. 818-864]. In a recent treatment of this subject [57], an equation is derived for the burning rate of a monopropellant droplet, where kinetics have been explicitly introduced, thereby eliminating the necessity of arbitrarily describing a combustion surface. In an attempt to apply these equations of droplet burning to conditions actually occurring in a rocket motor, Miesse [58] has considered the burning of a droplet when superimposed on a flow field, which is taken to be representative of that expected in a motor chamber. In addition to these works on individual droplets there have been several papers in which these concepts have been applied to the problem of spray combustion [59,60].

The effect of nonuniformity in the combustion pattern in a rocket motor has also been analyzed from a thermodynamic point of view [61]. It was found that for some systems the existence of variations in mixture ratio over the cross section can lead to a higher performance than if the mixture ratio were uniform. There have also been other studies on the observation of combustion in transparent motor chambers [62,63,64]. Although these observations are of qualitative interest, the process is too complex to draw quantitative relations from them. A problem which has recently received much attention in large scale rocket motors is that of instability during combustion [52, pp. 880-885; 65; 66; 67; 68; 69; 70, pp. 352-376]. In this connection consideration has been given to the "chugging" frequency in the motor chamber, acoustical vibrations, and mechanical feedback from the propellant pumping system.

L,13. Cited References.

1. Powell, W. B., and Kaplan, N. *Calif. Inst. Technol. Jet Propul. Lab. Rept. 24,* July 1944.
2. Kaplan, N., and Boden, R. H. *Calif. Inst. Technol. Jet Propul. Lab. Progr. Rept. 1-40,* Jan. 1946.
3. Bilfinger. Zündverzugsmessgerät für R-Stoffe. *B. M. W. Rept. 857,* Munich, Dec. 1944.
4. Broatch, J. D. *Fuel 29,* 106 (1950).
5. Gunn, S. V. *M.S. Thesis.* Purdue Univ., 1949.
6. Clark, F. B., and Sappington, M. H. *Calif. Inst. Technol. Jet Propul. Lab. Mem. 9-3,* June 1947.
7. Darling, B., and Kavanagh, G. M. *Mass. Inst. Technol. Rept. 19, Office of Nav. Research Contract N5-ori-78,* July 1947.
8. Collins, A. S. *Sc.D. Thesis.* Mass. Inst. Technol., 1948.
9. Gunn, S. V. *J. Am. Rocket Soc. 22,* 33 (1952).
10. Hirschfelder, J. O., and Curtiss, C. F. *J. Chem. Phys. 17,* 1076 (1950).
11. Bechert, C. *Portugaliae Phys. 3,* 29 (1949).
12. Zeldovich, Y. B. *Tech. Rept. F-TS-1226-IA (GDAM A9-T-45),* May 1949. (Transl. by Brown Univ.).
13. Rossini, F. D., Pitzer, K. S., Arnett, R. L., Braun, R. M., and Pimentel, G. C. *Selected Values of Physical and Thermodynamic Properties of Hydrocarbons and Related Compounds.* Carnegie Inst. Technol. Press, Pittsburgh, 1953.
14. Bichowsky, F. R., and Rossini, F. D. *The Thermochemistry of the Chemical Substances.* Reinhold, 1936.
15. Roth, W. A., and Isecke, K. *Ber. deut. chem. Ges. 77 B,* 537 (1944).
16. Holcomb, D. E., and Dorsey, C. L., Jr. *Ind. Eng. Chem. 41,* 2788 (1949).
17. Bellinger, F., Friedman, H. B., Bauer, W. H., Eastes, J. W., and Bull, W. C. *Ind. Eng. Chem. 40,* 1320, 1324 (1948).
18. Lothrop, W. C., and Handrick, G. R. *Chem. Revs. 44,* 419 (1949).
19. *Jet Propulsion.* Prepared by Staffs of Jet Propul. Lab. and Guggenheim Lab., Calif. Inst. Technol. for Air Tech. Service Command, 1946.
20. Steacie, E. W. R., and Folkins, H. O. *Can. J. Res. B17,* 105 (1939).
21. Mueller, K. H., and Walters, W. O. *J. Am. Chem. Soc. 73,* 1458 (1951).
22. Bloom, R., Davis, N. S., Jr., and Levine, S. D. *J. Am. Rocket Soc. 80,* 3 (1950).
23. Sanz, M. C. *J. Am. Rocket Soc. 75,* 122 (1948).
24. Kavanagh, G. M. *Mass. Inst. Technol. Rept. 28, D. I. C.,* 6552 (1949).
25. Williams, G. C., Satterfield, C. N., and Isbin, H. S. *J. Am. Rocket Soc. 22,* 70 (1952).
26. Davis, N. S., Jr., and Keefe, J. H., Jr. *J. Am. Rocket Soc. 22,* 63 (1952).
27. Thomas, D. D. *Calif. Inst. Technol. Jet Propul. Lab. Progr. Rept. 9-14,* 1947.
28. Altman, D., and Thomas, D. D. *Calif. Inst. Technol. Jet Propul. Lab. Progr. Rept. 9-36,* 1949.
29. Audrieth, L. F., et al. *N6-ori-71 Chemistry Task XX,* Chap. 1–10. Univ. Illinois, Oct. 1948–Feb. 1949.
30. Hager, K. F. *Ind. Eng. Chem. 41,* 2168 (1949).
31. Penner, S. S. *Calif. Inst. Technol. Jet Propul. Lab. Prog. Rept. 9-13,* Sept. 1947.
32. Hartwig, F. W. *A.E. Thesis.* Calif. Inst. Technol. Guggenheim Jet Propul. Center, 1952.
33. Spalding, D. B. *Fuel 29,* 2–7, 25–32 (1950).
34. Spalding, D. B. *Fuel 30,* 121–130 (1951).
35. Barr, J., and Mullins, B. P. Concerning combustion in vitiated atmospheres. *Natl. Gas Turbine Establishment (England) Rept. R44,* June 1949.
36. Godsave, G. A. E. The burning of single drops of fuel, Part I. *Natl. Gas Turbine Establishment (England) Rept. R66,* Mar. 1950.

37. Godsave, G. A. E.   The burning of single drops of fuel, Part II. *Natl. Gas Turbine Establishment (England) Rept. R87*, Apr. 1951.
38. Mullins, B. P.   Studies of the spontaneous ignition of fuels injected into a hot air stream, General Introduction and Part I. *Natl. Gas Turbine Establishment (England) Rept. R89*, Aug. 1951.
39. Mullins, B. P.   Studies on the spontaneous ignition of fuels injected into a hot air stream, Part II. *Natl. Gas Turbine Establishment (England) Rept. R90*, Aug. 1951.
40. Topps, J. E. C.   *J. Inst. Petroleum, 37*, 535–553 (1951).
41. Fieser, L. F., and Fieser, M.   *Organic Chemistry.* D. C. Heath, 1944.
42. Hughes, E. D., Ingold, C. K., et al.   *J. Chem. Soc.*, 2400–2473 (1950).
43. Ingold, C. K., Millen, D. J., et al.   Parts I–VI, *J. Chem. Soc.*, 2576–2627 (1950).
44. Sage, B. H., Hough, E. W., and Green, J.   *Calif. Inst. Technol. Jet Propul. Lab. Prog. Rept. 1-3*, Apr. 1942.
45. Gilman, H.   *Organic Chemistry*, Vol. I. Wiley, 1938.
46. Trent, C. H., and Zucrow, M. J.   *J. Am. Rocket Soc. 21*, 129 (1951).
47. Penner, S. S.   *J. Appl. Phys. 21*, 685–695 (1950).
48. Eyring, H., and Kassel, L. S.   *J. Am. Chem. Soc. 55*, 2796–2797 (1933).
49. Eyring, H.   *J. Am. Chem. Soc. 53*, 2537–2549 (1931).
50. Penner, S. S.   *J. Franklin Inst. 249*, 441–448 (1950).
51. Penner, S. S., and Datner, P. P.   *Tech. Rept. 7, Contract DA-495-Ord-446*, Sept. 1954. Presented at the *Fifth Symposium on Combustion*, Sept. 1954.
52. *Fourth Symposium on Combustion.* Williams & Wilkins, 1953.
53. Goldsmith, M., and Penner, S. S.   *Jet Propulsion 24*, 245–251 (1954).
54. Kumazai, S., and Isada, H.   *Science of Machine 4*, 337–342 (1952).
55. Graves, C. C.   Paper presented before the *Third Midwestern Conference on Fluid Mechanics*, Univ. Minn., 1953.
56. Wise, H., Lorell, J., and Wood, B. J.   Paper presented at the *Fifth Symposium on Combustion*, Sept. 1954.
57. Lorell, J., and Wise, H.   *Calif. Inst. Technol., Jet Propulsion Lab. Progress Rept. 20-237*, May 1954.
58. Miesse, C. C.   *J. Am. Rocket Soc. 24*, 237 (1954).
59. Probert, R. P.   *Phil. Mag. 37*, 94 (1946).
60. Graves, C. C., and Gerstein, M.   Paper presented before the *AGARD Combustion Panel*, Scheveningen, The Netherlands, May 1954. Butterworths, London (in press).
61. Altman, D., and Lorell, J.   *J. Am. Rocket Soc. 22*, 252–255 (1952).
62. Berman, K., and Cheney, S. H., Jr.   *J. Am. Rocket Soc. 23*, 89–96 (1953).
63. Berman, K., and Logan, S. E.   *J. Am. Rocket Soc. 22*, 78–85 (1952).
64. Altseimer, J. H.   *J. Am. Rocket Soc. 22*, 86–91 (1952).
65. Summerfield, M.   *J. Am. Rocket Soc. 21*, 108 (1951).
66. Crocco, L.   *J. Am. Rocket Soc. 21*, 163 (1951); *22*, 7–17 (1952).
67. Crocco, L., and Cheng, S. I.   *AGARD Combustion Instability.* Butterworths, London (in press).
68. Tsien, H. S.   *J. Am. Rocket Soc. 22*, 256 (1952).
69. Marble, F. E., and Cox, D. W., Jr.   *J. Am. Rocket Soc. 23*, 63 (1953).
70. Ross, C. C., and Datner, P. O.   *AGARD Selected Combustion Problems.* Butterworths, London, 1954.

SECTION G

THE LIQUID PROPELLANT ROCKET ENGINE

MARTIN SUMMERFIELD

G,1. Introduction. The scientific development of the modern rocket engine has been made possible to a great extent by the results of research in the fields of chemical physics, high temperature combustion, high intensity heat transfer, gas dynamics, and heat-resistant materials. Much of this work has taken place in recent years and has not been incorporated heretofore in the standard textbook literature in a form suitable for systematic study. For an up-to-date comprehensive treatment of the liquid propellant rocket engine, therefore, it would have been necessary to introduce in this section all of the new developments mentioned. However, in the planning of this series, many of these topics were placed in other sections to allow a more logical arrangement of the entire subject matter. Therefore, in this section, the pertinent results or conclusions derived from these modern investigations are merely stated in their simplest form, and the reader is directed by cross references to the other volumes where the complete treatments can be found.

In rocketry, as in any field that has grown rapidly, the terminology has not yet been fully standardized and accepted. Therefore it is appropriate to start with a few definitions.

Rocket propulsion is a system of propulsion that depends on forward thrust created by rearward ejection of a fluid jet through a nozzle mounted in the vehicle, with the special condition that the fluid in the jet originate entirely from tanks within the vehicle. It is this special condition that distinguishes the rocket from other classes of jet engines that ingest the surrounding medium (air or water) to form the driving jet. Therefore the rocket engine is able to function not only under the usual conditions of flight through the atmosphere (or under water), but in the vacuum outside the atmosphere.

A *rocket propellant* is the fluid substance that forms the driving jet, although the term is used most frequently to refer to the driving fluid in its chemical state before combustion. The term is used also to denote *one* of the reactants in a multicomponent propellant system. A *fuel* is any propellant that can burn in the presence of oxygen, and the term includes not only hydrocarbons but other substances (e.g. ammonia, powdered aluminum) as well. An *oxidizer* is a propellant that can support the combustion of a fuel, and is applicable to substances that may not contain oxygen (e.g. fluorine) as well as those that do. A *bipropellant system* is one that consists of two reactants, usually a fuel and an oxidizer. A *monopropellant* is a single substance that can be caused to react in the combustion chamber to generate hot gas to form the driving jet. This term applies strictly only to a single compound (e.g. ethylene oxide) that undergoes a decomposition reaction, but it has been applied also to a propellant mixture that is stored in a single propellant tank (e.g. a mixture of methyl nitrate and methyl alcohol).

The *rocket motor* is usually understood to be the part of the engine in which the propellants are burned and the jet is formed, while the term *rocket engine* usually refers to the entire propulsion system, including the tanks if they are constructed integrally with the engine. The term *thrust cylinder* has been used in place of rocket motor in some writings, but the latter term is historically the oldest and the most widely preferred. In the conventional solid propellant rocket, the engine and the motor happen to be the same piece of apparatus because the propellant is stored in the combustion chamber of the rocket motor; this is not so in the liquid propellant rocket.

G,2. Performance Analysis of the Ideal Rocket Motor. The performance analysis of a rocket motor comprises calculations of the thrust F, the effective exhaust velocity c, the adiabatic combustion temperature in the chamber T_c, the thrust coefficient C_F, the characteristic velocity c^*, and certain efficiencies η. Performance parameters derived from c are the specific impulse I_{sp} and the specific propellant consumption w_{sp}.

The thrust equation is the fundamental starting point. In general, the thrust exerted on a duct of arbitrary shape can be calculated from the momentum equation written in integrated form appropriate for one-dimensional flow problems. (See III,B.) Let $\dot{m}_i$ be the rate of mass flow into the inlet, p_i the static pressure at the inlet, V_i the stream velocity at the inlet, and A_i the area of the inlet, and let the corresponding quantities at the exit be indicated by the subscript e. Then,

$$\text{The stream thrust at the inlet} = (\dot{m}_i V_i + p_i A_i)$$

$$\text{The stream thrust at the exit} = (\dot{m}_e V_e + p_e A_e)$$

$$\text{The total force on the external duct surface} = F + p_\infty (A_e - A_i)$$

$$F + p_\infty (A_e - A_i) = (\dot{m}_e V_e + p_e A_e)$$
$$- (\dot{m}_i w_i + p_i A_i)$$

The external force on the duct is expressed here as if the pressure on the external surface were identical with the ambient pressure p_∞ of the atmosphere, although in actual flight this is not so. Therefore, for a duct in flight, this equation implies a certain arbitrary separation between the thrust F and the aerodynamic drag D. Separation in this manner is justified by its convenience, because the thrust measured in a ground test of the propulsion system is closely equal to the thrust F thus calculated.

In the particular case of a rocket, $\dot{m}_i$ and A_i may be set equal to zero. Then,

$$F = \dot{m}V_e + (p_e - p_\infty)A_e \qquad (2\text{-}1)$$

With reasonably well-designed exhaust nozzles, the exit pressure p_e is nearly or exactly equal to the ambient pressure p_∞, so that the second term is in the nature of a small correction to the thrust. This makes it convenient to define an *effective exhaust velocity* c such that

$$F = \dot{m}c \qquad (2\text{-}2)$$

$$c = V_e + \frac{(p_e - p_\infty)A_e}{\dot{m}} \qquad (2\text{-}3)$$

Clearly, c equals V_e if the nozzle is designed properly, that is, if the p_e equals p_∞. For any given values of chamber pressure p_c and of p_∞ and $\dot{m}$, both F and c reach their maximum values when the exit area A_e of the nozzle is chosen to produce a static pressure p_e at the exit exactly equal to p_∞. (This will be proved later.) The exit velocity and the correction term $(p_e - p_\infty)A_e/\dot{m}$ both vary strongly with A_e, but in opposite directions so that the sum is quite insensitive to A_e, that is, the maximum is very flat. As a result, the effective exhaust velocity measured by the ratio of F to $\dot{m}$ can be taken to be the value of $(V_e)_{opt}$, even if the actual nozzle used in the test is somewhat off-design. Herein lies the practical significance of the concept of the effective exhaust velocity (see Fig. G,2a.)

The specific propellant consumption w_{sp}, defined as the weight rate of consumption of propellant $\dot{w}$ per unit thrust, is another useful index of rocket engine performance. Let g_0 denote the standard acceleration due to gravity.

$$w_{sp} = \frac{\dot{w}}{F} = \frac{\dot{m}g_0}{F} = \frac{g_0}{c} \qquad (2\text{-}4)$$

The specific impulse I_{sp} (called specific thrust in some writings) is defined as the propulsive impulse delivered by the engine per unit weight of propellant.

$$I_{sp} = \frac{Fdt}{dw} = \frac{F}{\dot{w}} = \frac{1}{w_{sp}} = \frac{c}{g_0} \qquad (2\text{-}5)$$

A figure of merit that is sometimes quoted is the impulse-weight ratio, I/W, of a loaded rocket propulsion system or of a loaded rocket vehicle.

If the firing program calls for constant pressure and hence constant specific impulse,

$$\frac{I}{W} = I_{sp}\nu \qquad (2\text{-}6)$$

where ν is the *propellant loading fraction*, that is, the ratio of the initial mass of propellant when the rocket is fully loaded to the gross mass of the loaded rocket. Obviously, ν approaches unity as the structural effectiveness is improved. Therefore, I/W measures both the performance of the engine and the effectiveness of the structure of the rocket.

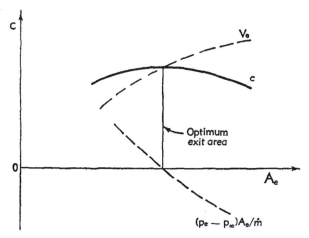

Fig. G,2a. Variation of effective exhaust velocity with exit area.

The matter of units of specific impulse deserves comment. The definition involves the mass rate of flow of propellant expressed in weight units, so that in the fps system I_{sp} should be lb sec/lb. It is common practice to denote this ratio simply as sec, canceling the lb in numerator and denominator, disregarding the fact that one lb refers to a force and the other to a mass. That this is an error becomes apparent when it is realized that the specific impulse has the dimensions of a velocity.

In order to develop several additional performance parameters, it is necessary to describe in detail the thermodynamic and gas dynamic processes in the rocket motor. As a starting point, it is a great simplification to deal with the so-called *ideal rocket motor*. From a practical standpoint, the ideal rocket motor is a useful concept because it leads to simple theoretical formulas for F, c, T_c, C_F, and c^*, which otherwise would have to be presented in tables or in graphs.

The ideal rocket motor analysis rests on the following simplifications: (1) the propellant gas obeys the perfect gas law; (2) its specific heat is constant, independent of temperature; (3) the flow is parallel to the axis of the motor and uniform in every plane normal to the axis, thus constituting a one-dimensional problem; (4) there is no frictional dissipation in the chamber or nozzle; (5) there is no heat transfer to the motor walls; (6) the flow velocity in the chamber before the nozzle entrance is zero; (7) combustion or heat addition is completed in the chamber at *constant pressure and does not occur in the nozzle;* and (8) the process is steady in time.

The thermodynamic process is indicated in Fig. G,2b, both in the p, V diagram and in the h, s diagram. Combustion at constant pressure

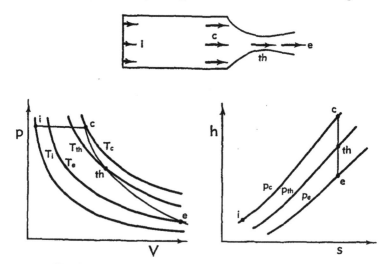

‘Fig. G,2b. Ideal thermodynamic processes in the combustion chamber and nozzle of a rocket motor.

moves the state point from i to c. The highest temperature occurs at c. Then, since the frictionless and adiabatic conditions assumed for the ideal rocket motor imply an isentropic expansion in the nozzle (see III,B for proof; also I,A), the state point moves along the constant entropy line from c to e during the expansion process. As indicated in the motor sketch of Fig. G,2b, the subscripts i, c, th, and e employed in the following analysis refer respectively to the unburned state at the injector, the all-burned state in the chamber just before expansion, the state at the throat of the de Laval nozzle, and the state at the exit of the nozzle.

The combustion temperature (or adiabatic flame temperature) T_c is determined by the heat of combustion at constant pressure per unit mass Δh_c.

$$\Delta h_c = c_p(T_c - T_i) \tag{2-7}$$

At any station in the nozzle, the entropy, pressure, temperature, velocity, and Mach number are given by the following relations (III,B):

$$p = \rho \frac{R}{\mathfrak{M}} T; \quad c_p - c_v = \frac{R}{\mathfrak{M}}; \quad c_p = \frac{\gamma}{\gamma - 1} \frac{R}{\mathfrak{M}} \tag{2-8}$$

$$s = s_o; \quad \frac{T}{T_o} = \left(\frac{p}{p_o}\right)^{\frac{\gamma-1}{\gamma}} \tag{2-9}$$

$$\left. \begin{aligned} \tfrac{1}{2}V^2 &= c_p(T_o - T) = \frac{\gamma}{\gamma - 1} \frac{RT_o}{\mathfrak{M}} \left[1 - \left(\frac{p}{p_o}\right)^{\frac{\gamma-1}{\gamma}}\right] \\[2mm] \frac{A}{A_{th}} &= \frac{1}{M} \left[\frac{1 + \dfrac{\gamma - 1}{2} M^2}{\dfrac{\gamma + 1}{2}}\right]^{\frac{\gamma+1}{2(\gamma-1)}} \end{aligned} \right\} \tag{2-10}$$

Since the over-all pressure ratio p_o/p_∞ is always sufficiently large in rockets to establish sonic flow at the throat, then

$$M_{th} = \frac{V_{th}}{a_{th}} = 1; \quad V_{th} = \left(\gamma \frac{R}{\mathfrak{M}} T_{th}\right)^{\frac{1}{2}} \tag{2-11}$$

$$\frac{p_{th}}{p_o} = \left(\frac{2}{\gamma + 1}\right)^{\frac{\gamma}{\gamma-1}}; \quad \frac{T_{th}}{T_o} = \frac{2}{\gamma + 1} \tag{2-12}$$

The specific heat ratios of typical rocket jet gases range between 1.1 and 1.3. The first figure corresponds to mixtures at very high temperatures with large concentrations of water vapor and large effective specific heats due to strong dissociation; the latter figure applies to moderate temperature gases with moderate concentrations of H_2O and CO_2. With $\gamma = 1.2$, it can be seen that the drop in pressure from the chamber to the throat is nearly half the chamber pressure, while the drop in temperature is only about one tenth the chamber temperature.

The mass flow through the nozzle can be expressed in terms of the flow conditions at any station. Let A be the cross-sectional area at any station:

$$\dot{m} = \rho V A = p_o A \left\{\frac{2\gamma}{\gamma - 1} \frac{\mathfrak{M}}{RT_o} \left(\frac{p}{p_o}\right)^2 \left[1 - \left(\frac{p}{p_o}\right)^{\frac{\gamma-1}{\gamma}}\right]\right\}^{\frac{1}{2}} \tag{2-13}$$

A plot of mass flow per unit area $\dot{m}/A$ against static pressure ratio p/p_o, shown in Fig. G,2c, exhibits a maximum at the throat of the de Laval nozzle, just as expected.

$$\frac{\dot{m}}{A_{th}} = p_o \left\{\gamma \left(\frac{2}{\gamma + 1}\right)^{\frac{\gamma+1}{\gamma-1}} \frac{\mathfrak{M}}{RT_o}\right\}^{\frac{1}{2}} \tag{2-14}$$

By equating the two expressions, Eq. 2-13 and 2-14, for $\dot{m}$, an expression for A/A_{th} is obtained.

$$\frac{A}{A_{th}} = \frac{\left(\dfrac{\gamma-1}{2}\right)^{\frac{1}{2}} \left(\dfrac{2}{\gamma+1}\right)^{\frac{\gamma+1}{2(\gamma-1)}}}{\left(\dfrac{p}{p_o}\right)^{\frac{1}{\gamma}} \left[1 - \left(\dfrac{p}{p_o}\right)^{\frac{\gamma-1}{\gamma}}\right]^{\frac{1}{2}}} \tag{2-15}$$

By inserting A_e and p_e for A and p in Eq. 2-15, the nozzle area ratio $\epsilon \; (\equiv A_e/A_{th})$ can be expressed as a function of p_o/p_e. This relation is plotted in Fig. G,2d for several values of γ.

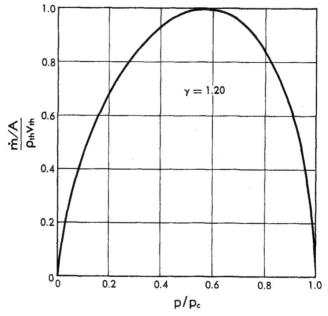

Fig. G,2c.   Variation of mass flow per unit area $\dot{m}/A$ with pressure ratio p/p_o.

The maximum exit velocity $(V_e)_{max}$, obtained by setting p_e/p_o in Eq. 2-10 equal to zero, is

$$(V_e)_{max} = \sqrt{\frac{2\gamma}{\gamma-1}\frac{RT_e}{\mathfrak{M}}} \tag{2-16}$$

It is interesting that the maximum exit velocity, obtained by expansion to zero pressure, is greater than the root-mean-square molecular velocity in the chamber by the factor $[2\gamma/3(\gamma-1)]^{\frac{1}{2}}$, or about 2 for $\gamma = 1.2$. This result follows directly, of course, from molecular energy considerations.

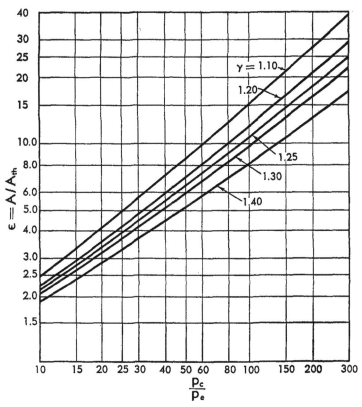

Fig. G,2d.   Variation of nozzle area ratio with pressure ratio. After Sutton,
G.P., *Rocket Propulsion Elements*, Wiley, 1956, by permission.

The thrust formula (Eq. 2-1) can now be expressed in terms of the
pressures by substituting Eq. 2-10 for V_e and Eq. 2-14 for $\dot{m}$.

$$F = p_c A_{th} \left\{ \frac{2\gamma^2}{\gamma - 1} \left(\frac{2}{\gamma + 1} \right)^{\frac{\gamma+1}{\gamma-1}} \left[1 - \left(\frac{p_e}{p_c} \right)^{\frac{\gamma-1}{\gamma}} \right] \right\}^{\frac{1}{2}} + (p_e - p_\infty) A_e \quad (2\text{-}17)$$

From this formula, it is clear that the thrust does not depend at all
on the combustion temperature T_c, but depends mainly on the dimensions
of the nozzle A_e and A_{th}, and on the chamber pressure p_c.

An important performance parameter, the rocket thrust coefficient
C_F, can now be deduced. The defining equation is

$$C_F \equiv \frac{F}{p_c A_{th}} \quad (2\text{-}18)$$

From Eq. 2-17 C_F can be evaluated:

$$C_F = \left\{ \frac{2\gamma^2}{\gamma - 1} \left(\frac{2}{\gamma + 1} \right)^{\frac{\gamma+1}{\gamma-1}} \left[1 - \left(\frac{p_e}{p_c} \right)^{\frac{\gamma-1}{\gamma}} \right] \right\}^{\frac{1}{2}} + \left(\frac{p_e}{p_c} - \frac{p_\infty}{p_c} \right) \epsilon \quad (2\text{-}19)$$

Since p_e/p_0 is a function of ϵ according to Eq. 2-15, C_F depends only on the three independent variables γ, p_c/p_∞, and ϵ. Graphs of this function are presented in Fig. G,2e and G,2f for $\gamma = 1.2$ and $\gamma = 1.3$, respectively.

There are several features of these curves that deserve attention. First, each curve shows a maximum value of C_F at a certain area ratio which may, for this purpose, be called ϵ_{opt}. It can be shown analytically, by differentiating Eq. 2-19 and setting the derivative equal to zero, that the peak value occurs for a value of ϵ such that $p_e = p_\infty$, that is, for a

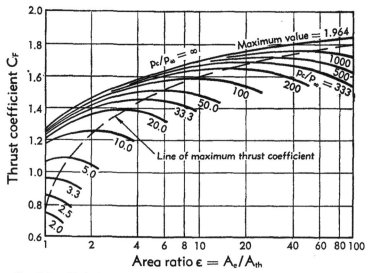

Fig. G,2e. Variation of rocket thrust coefficient with nozzle area ratio and pressure ratio p_c/p_∞ for $\gamma = 1.2$. After Sutton, G.P., *op. cit.*

properly expanded nozzle. The area ratio ϵ_{opt} for proper expansion can be determined from the peak of the appropriate curve in Fig. G,2e or G,2f, or, more accurately, from Fig. G,2d. A nozzle having an area ratio less than ϵ_{opt} is said to be *underexpanded*, and one having an area ratio more than ϵ_{opt} is *overexpanded*. Clearly, nozzles that are either overexpanded or underexpanded produce less thrust than a properly expanded nozzle. This conclusion can be proved in another way, by considering the distribution of pressure on the inner and outer surfaces of the rocket motor, as shown in Fig. G,2g. Downstream of the section indicated as ϵ_{opt}, the internal pressure is less than the external pressure, so that this portion of the cone acts to produce a force opposed to the thrust of the rocket motor as a whole. Therefore it is better to dispense with it and to terminate the nozzle at ϵ_{opt}. Consequently the highest thrust is produced with a properly expanded nozzle.

Examination of the pressure distribution pictured in Fig. G,2f shows that C_F must be somewhat greater than unity, except for the unusual case of low chamber pressure $(p_0/p_0 \cong 1)$ and for possibly greatly overexpanded nozzles. Viewed in the simplest way, the rocket motor is a pressurized vessel with a hole of area A_{th} in the aft wall, and so it would be

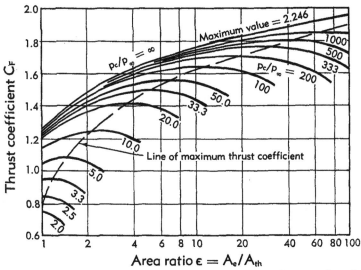

Fig. G,2f.   Variation of rocket thrust coefficient with nozzle area ratio and pressure ratio p_0/p_∞ for $\gamma = 1.3$. After Sutton, G.P., *op. cit.*

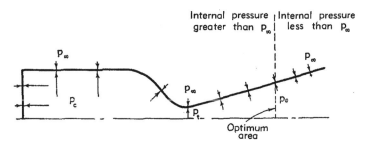

Fig. G,2g.   Optimization of nozzle area ratios.

acted upon by an unbalanced force $(p_c - p_0)A_{th}$. To this must be added the additional unbalance due to the pressure reduction in the entrance cone of the nozzle. Then, if $p_0 \ll p_c$, the ratio $F/p_c A_{th}$ must be somewhat larger than 1.

The maximum values of C_F on the families of curves in Fig. G,2e and G,2f can be connected by a smooth curve obtained by setting $p_e = p_\infty$

in Eq. 2-19.

$$(C_F)_{max} = \left\{ \frac{2\gamma^2}{\gamma - 1} \left(\frac{2}{\gamma + 1} \right)^{\frac{\gamma+1}{\gamma-1}} \left[1 - \left(\frac{p_\infty}{p_o} \right)^{\frac{\gamma-1}{\gamma}} \right] \right\}^{\frac{1}{2}} \qquad (2\text{-}20)$$

The dependence on (p_∞/p_o) can be replaced by dependence on ϵ_{opt} through the relation (Eq. 2-15). The curve $(C_F)_{max}$ vs. ϵ_{opt} itself reaches an ultimate value for infinite expansion. Thus, if (p_∞/p_o) is set equal to zero in Eq. 2-20,

$$(C_F)_{ult} = \left[\frac{2\gamma^2}{\gamma - 1} \left(\frac{2}{\gamma + 1} \right)^{\frac{\gamma+1}{\gamma-1}} \right]^{\frac{1}{2}} \qquad (2\text{-}21)$$

For example, $(C_F)_{ult} = 2.246$ for $\gamma = 1.2$.

Inspection of Fig. G,2e and G,2f discloses that, for a prescribed nozzle area ratio ϵ, the thrust coefficient increases monotonically as p_o/p_∞ increases. This becomes clear when Eq. 2-19 is written in terms of the thrust coefficient for a given nozzle when it is operating in a vacuum:

$$C_F = (C_F)_{vac} - \epsilon \frac{p_\infty}{p_o} \qquad (2\text{-}22)$$

Thus, for given ϵ and p_o, the increase in thrust with increasing p_o/p_∞ stems entirely from the reduction of the pressure acting on the external surfaces of the rocket motor.

It is significant that C_F is completely independent of combustion temperature T_c and of molecular weight $\mathfrak{M}$. Consequently, as a figure of merit, it is insensitive to the efficiency of combustion, but it is sensitive to the quality of the exhaust nozzle. In practice, the test engineer compares the measured C_F, computed from p_o, A_{th}, and F by means of Eq. 2-18, with the theoretical C_F computed from Eq. 2-19 to determine whether the nozzle is functioning properly, and in this way he can localize to some extent the cause of an unexpected defect in specific impulse. The other possible area for loss is in the combustion process. To detect combustion inefficiency, the performance parameter c^* is useful.

The characteristic velocity c^* is defined as follows:

$$c^* \equiv \frac{p_o A_{th}}{\dot{m}} \qquad (2\text{-}23)$$

It follows immediately from Eq. 2-2 and 2-18 that

$$c = C_F c^* \qquad (2\text{-}23a)$$

A theoretical expression for c^* is obtainable from Eq. 2-14.

$$c^* = \left[\frac{1}{\gamma} \left(\frac{\gamma + 1}{2} \right)^{\frac{\gamma+1}{\gamma-1}} \frac{R T_c}{\mathfrak{M}} \right]^{\frac{1}{2}} \qquad (2\text{-}24)$$

From this formula it appears that c^* depends mainly on conditions in the combustion chamber, that is, on flame temperature and combustion product composition. Consequently, just as C_F is used as an index of the quality of the exhaust nozzle, so c^* is used in practice as an index of the efficiency of combustion. The test engineer determines c^* from measured values of p_c, A_{th}, $\dot{m}$, and compares it with the theoretical value (Eq. 2-24). In this way, a defect in specific impulse can be traced to a possible loss in the combustion process. Although the performance of a rocket motor is adequately described by the exhaust velocity c, which requires only the measurement of F and $\dot{m}$, it is the usual practice to measure at the same time p_c and A_{th} in order to compute C_F and c^* for diagnostic purposes.

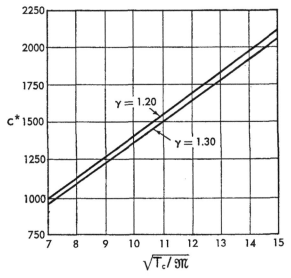

Fig. G,2h.   Variation of characteristic velocity c^* with $\sqrt{T_c/\mathfrak{M}}$.

(Careful consideration of the flow process reveals that there exists some slight cross-dependence of C_F and c^*, that is, the former is slightly sensitive to the combustion process and the latter is somewhat affected by the flow conditions in the nozzle, but this is usually ignored.)

Curves of c^* vs. $(T_c/\mathfrak{M})^{\frac{1}{2}}$ for $\gamma = 1.2$ and $\gamma = 1.3$ are plotted in Fig. G,2h. It is significant that c^*, and therefore the specific impulse, depends as much on molecular weight $\mathfrak{M}$ as on flame temperature. Thus, it is just as important for the product gas to have a low mean molecular weight as a high temperature. This point will arise later when actual propellants are considered. It will be pointed out then that the optimum fuel-oxidizer mixture ratio is not necessarily the one that produces the highest flame temperature, and that a particular propellant combination may be very hot but no better than another much cooler one from the standpoint of specific impulse.

The magnitude of c^* is of interest. Because of its dependence on $(RT_c/\mathfrak{M})^{\frac{1}{2}}$, it can be compared directly with the velocity at the nozzle throat. Thus, for $\gamma = 1.2$, $c^* = 1.5V_{th}$. In general, for a properly designed nozzle, $V_{th} < c^* < V_e$, so that c^* equals the gas velocity at some station in the divergent part of the exhaust nozzle.

The efficiency of the rocket engine can be discussed in terms of the concepts of the ideal rocket motor. Five efficiencies deserve discussion here: combustion efficiency, expansion or "cycle" efficiency, nozzle efficiency, thermal efficiency, and total efficiency. In addition, the attempts to define a so-called propulsive efficiency will be examined.

The *combustion efficiency* η_c is defined as the ratio of the actual enthalpy released by combustion to the ideal enthalpy that would be released if the reaction were to go to completion.

$$\eta_c = \frac{(\Delta h_c + h_i)_{actual}}{(\Delta h_c + h_i)_{ideal}} = \frac{(T_c)_{actual}}{(T_c)_{ideal}} \tag{2-25}$$

The *ideal expansion or cycle efficiency* η_i expresses the fraction of the enthalpy available in the combustion chamber that can ideally be converted to kinetic energy in the exhaust jet. Let h_i represent the enthalpy of the product gas at the injection temperature with reference to $0°K$.

$$\eta_i = \frac{(\frac{1}{2}V_e^2)_{ideal}}{\Delta h_c + h_i} = \frac{(\frac{1}{2}V_e^2)_{ideal}}{c_p T_c} \tag{2-26}$$

From Eq. 2-10,

$$\eta_i = 1 - \left(\frac{p_\infty}{p_c}\right)^{\frac{\gamma-1}{\gamma}} \tag{2-27}$$

The *nozzle efficiency* η_n is defined as the ratio of the actual kinetic energy in the exhaust jet to that which could be produced ideally at the specified pressure ratio. (Compare with diffuser efficiency in ramjets, Sec. E.)

$$\eta_n = \frac{(\frac{1}{2}V_e^2)_{actual}}{(\frac{1}{2}V_e^2)_{ideal}} = \frac{(T_c - T_e)_{actual}}{(T_c - T_e)_{ideal}} \tag{2-28}$$

The *thermal efficiency* η_{th} is the ratio of actual kinetic energy in the exhaust jet to the total enthalpy that could ideally be produced by the combustion reaction.

$$\eta_{th} = \frac{(\frac{1}{2}V_e^2)_{actual}}{(\Delta h_c)_{ideal} + h_i} \tag{2-29}$$

Conversion losses in the combustion process and in the expansion process, and the enthalpy discarded in the hot exhaust jet, are all represented in the thermal efficiency.

$$\eta_{th} = \eta_c \eta_n \eta_i = \frac{(V_e^2)_{actual}}{(V_e^2)_{max}} \tag{2-30}$$

The four efficiencies mentioned so far refer to the rocket motor simply as a heat engine and do not involve the energy quantities connected with flight. When flight is considered, it is possible to define an *over-all* or a *total efficiency* η_0 as the ratio of propulsive power (output) to the rate of consumption of energy in the rocket motor (input). The output is simply $V_\infty V_e$ per unit mass flow of propellant, where V_∞ is the flight speed. The input is the sum of $(\Delta h_c + h_i)$ plus the kinetic energy $\frac{1}{2}V_\infty^2$ of the propellant. The latter quantity results from the recognition that a quantity of propellant possesses more total energy when it is in motion than when at rest.

$$\eta_0 = \frac{V_\infty V_e}{(\Delta h_c + h_i)_{ideal} + \frac{1}{2}V_\infty^2}$$

$$= \sqrt{\eta_{th}} \left[\frac{2V_\infty (V_e)_{max}}{(V_e^2)_{max} + V_\infty^2} \right]$$

(2-31)

The maximum value of η_0 is $\sqrt{\eta_{th}}$, and this value is reached when the flight velocity equals the maximum theoretical exhaust velocity. (This result has led to statements in the early literature that a rocket's flight speed could not exceed the maximum theoretical exhaust velocity. Clearly, this conclusion is not justified.)

There have been several attempts in the past to define a suitable "propulsive efficiency" by analogy with the corresponding case of a propeller-driven airplane, but these attempts have always failed because of the lack of a logical definition of mechanical input. If the analogy is to be carried through, the so-called propulsive efficiency will have to satisfy three requirements: (1) it should be a ratio of a mechanical output to a mechanical input; (2) the product of the propulsive efficiency and the thermal efficiency η_{th} should be equal to the total efficiency η_0; and (3) it must under no flight condition exceed unity.

The definition of propulsive efficiency that has been most prominent in the literature is

$$\eta_p = \frac{output}{input} = \frac{output}{output + loss} = \frac{V_\infty V_e}{V_\infty V_e + \frac{1}{2}(V_e - V_\infty)^2} = \frac{2V_\infty V_e}{(V_\infty^2 + V_e^2)}$$

The denominator is supposed to represent the sum of the propulsive work and the absolute kinetic energy of the jet (a loss). The present author objects to this propulsive efficiency on the grounds that it offers no clear definition of mechanical input and that, even if this particular definition were allowed, it fails to meet the requirement that $\eta_p \eta_{th} = \eta_0$.[1]

As a conclusion to this discussion of efficiency, it may be remarked that the concept of efficiency has not been at all useful in the field of rocket engines. Rockets are always compared on the basis of I_{sp} or c^* values, or on specific propellant consumption, and rarely is efficiency

[1] Editor's note: The author would find the same difficulty with any engine whenever the kinetic energy of the fuel is included.

mentioned. This is logical since flight performance depends directly on I_{sp}, and only indirectly on η.

Consideration of the thermal efficiency leads to the idea of a heat balance for the rocket motor. The chemical energy introduced with the injected propellants is distributed in four directions: (1) a small part (up to 15 per cent) remains unconverted due to incomplete reaction; (2) about 1 per cent, more or less, of the heat reaction is transferred to the motor walls where it may be lost unless the motor is cooled by the liquid propellant (regenerative cooling); (3) the kinetic energy of the jet comprises a large part, from one third to two thirds, of the heat of reaction; and (4) finally, the remainder of the reaction heat is carried away in the hot jet as thermal energy.

Typical values of the performance parameters of rocket motors in the moderate and high performance classes, respectively, are indicated in Table G,2. It is notable that the flame temperatures range from 2000

Table G,2. Typical performance characteristics.

Performance parameter	Ordinary range	High range
T_c	2000–3000°K	3000–5000°K
c^*	4000–5500 ft/sec	5000–8000 ft/sec
C_F	1.3–1.5	1.5–1.6
I_{sp}	200–270 lb-sec/lb	270–400 lb-sec/lb
$\overline{\mathfrak{M}}$	20–25	8–20
γ	1.15–1.25	1.15–1.20

to 5000°K, and that dissociation is prominent in most of the range. The effects of dissociation constitute the greatest source of inaccuracy in the application of the ideal analysis just presented to actual rocket processes. The modification of the ideal analysis to handle dissociation will be discussed in Art. 4. Modifications due to other departures from the ideal conditions of this analysis are discussed in the following article.

G,3. Departures from Ideal Performance. The preceding ideal performance analysis requires correction to take care of the following actual conditions:

a. Conical divergence of the exhaust jet.
b. Surface friction and flow disturbances in the exhaust nozzle.
c. Constriction of the exit area due to boundary layer build-up.
d. Jet detachment.
e. Heat loss from the hot gas to the cold motor walls.
f. Suspended liquid or solid particles in the exhaust jet.
g. Pressure drop in the combustion chamber due to heat release.

Consideration of the effects of temperature dependence of the specific heats and of chemical reaction during the expansion in the nozzle will be postponed until Art. 4.

The effect of exhaust nozzle divergence. A loss in thrust occurs as a result of the divergence of the exhaust jet as it leaves the nozzle, in comparison with the ideal case of parallel flow. The loss may be viewed as the decrease in the axial component of momentum due to the outward inclination of the streamlines. Inasmuch as the loss is usually only a few per cent, a fairly approximate theory is adequate to determine its magnitude.

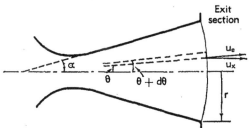

Fig. G,3a.   Radical flow in exhaust nozzle.

If the exit-to-throat area ratio ϵ is sufficiently large, say at least 3, and if the half angle α of the exit cone is sufficiently small, say not more than 20°, it is reasonable to assume a radial flow pattern in the exhaust cone with concentric spherical caps for isobaric surfaces. The exit cone defines the location of the effective center for this radial flow pattern. The axis of the nozzle corresponds to $\theta = 0$, and the surface of the exit cone corresponds to $\theta = \alpha$ (Fig. G,3a).

To calculate the thrust, divide the exit flow area into differential annular areas between θ and $\theta + d\theta$. The axial component of the thrust of this element of flow is

$$dF = \rho_e V_e (2\pi r \sin \theta \cdot r d\theta)(V_e \cos \theta) + (p_e - p_\infty)(2\pi r \sin \theta \cdot r d\theta) \cos \theta$$

Integrating with respect to θ from 0 to α, F becomes

$$F_\alpha = \left(\frac{1 + \cos \alpha}{2} \right) [\dot{m} V_e + (p_e - p_\infty) A_e'] \tag{3-1}$$

In this formula, V_e, p_e, and A_e' come out in the analysis to be the velocity, pressure, and area of the *spherical isobaric cap* at the exit, and not the planar area at the exit. The quantity in brackets is the thrust of a zero divergence motor whose exit area equals A_e'. However, for small angle α, the planar exit area A_e is so close to A_e' that the thrust of the zero divergence motor would be almost unaffected if A_e' were replaced by A_e, and V_e and p_e were interpreted to apply to A_e instead of

A'_e. Therefore, if F_0 is the thrust of a zero divergence motor having an exit area ratio ϵ equal to the *planar* exit area ratio of the α divergence motor,

$$F_\alpha = \lambda F_0, \quad \text{where} \quad \lambda = \tfrac{1}{2}(1 + \cos \alpha) \tag{3-2}$$

For $\alpha = 10°$, $\lambda = 0.992$, and for $\alpha = 20°$, $\lambda = 0.970$. Consequently, the correction is a small one, and the preceding simplified model for the flow field is justified. To measure the loss accurately in order to check the theory requires experiments of high precision. Some experimental results have been reported, which show that the formula (Eq. 3-2) holds reasonably well from $\alpha = 10°$ to $\alpha = 30°$. For smaller angles, the nozzles are so long that frictional losses dominate the results, and for angles in excess of $40°$ there is clear evidence of flow separation.

The most commonly employed divergence angle in practice is about $15°$. This produces a loss of only 1.7 per cent, yet the nozzle is not so long as to be excessive in either weight or length. It is possible to design an exit section with a curved contour so as to produce a parallel jet and thereby recover this divergence loss, but the penalties of complexity and cost must then be accepted. For a design to produce the utmost in performance, e.g., a satellite launcher, this is worthwhile.

The effects of fluid friction and flow losses. It is not possible to calculate reliably the thrust loss due to surface friction, because so little is known of the nature of the boundary layer under the conditions of pressure gradient and curvature that exist in the nozzle. Also, the flow field in a practical nozzle is likely to exhibit strong pressure waves originating in the sharply curved region near the throat. The rocket designer ordinarily cannot eliminate such sharp curvature, as can the supersonic wind tunnel designer, because rocket nozzles must be made cheaply (hence simply) and must be compact. Consequently, the flow disturbances are accepted as long as the loss in thrust is small. Fortunately, it seems that drastically poor design is necessary to cause a thrust loss greater than, say 10 per cent, due to bad flow, and if the designer merely uses his French curves with good technique, the loss can be cut to about 1 or 2 per cent.

For example, if the nozzle contour consists of a converging cone of $30°$ half angle, a diverging cone of $15°$ half angle, and a throat section with a longitudinal radius of curvature equal to twice the radius of curvature of the throat section, the loss will ordinarily not exceed 3 per cent and will probably be nearer to 1 per cent if there are no sharp corners.

In test results, this type of loss is reported in the form of a discharge coefficient C_d that normally lies between 0.97 and 0.99.

$$(C_F)_{actual} = C_d \lambda [(C_F)_{\alpha=0}]_{ideal} \tag{3-3}$$

Constriction of the exit area by the boundary layer. The effect of the boundary layer in constricting the flow passage of the nozzle can be considered in two places, at the throat and at the exit. At the throat, a boundary layer would reduce the mass flow $\dot{m}$; at the exit, it would reduce the effective expansion by reducing ϵ. It turns out that the first effect is negligible and the second one is small.

Observations of boundary layer growth in supersonic wind tunnels indicate that, in the Reynolds number range from 5×10^5 to 5×10^6, and in the Mach number range from 1.5 to 2.5, the boundary layer is turbulent and the displacement thickness grows in proportion to the distance from the throat. This Reynolds number is based on the velocity and kinematic viscosity at the particular station and the distance from the throat to this station. Although the strong cooling of the boundary layer and the much larger pressure gradient in the rocket nozzle would suggest caution in applying wind tunnel results to rocket nozzles, there are some observations in cold flow through rocket-type nozzles which indicate similar behavior. (This conclusion will be utilized in Art. 5 in the discussion of convective heat transfer in rocket nozzles.)

$$\delta^* \text{ (displacement thickness)} \cong 0.004 l_n \text{ (distance from throat)} \quad (3\text{-}4)$$

With this relation it turns out that the exit area constriction for a 15° half angle nozzle is only 3 per cent. Inspection of Fig. G,2e and G,2f shows that the effect of a 3 per cent area reduction can be neglected if the nozzle area ratio is near the optimum. For narrower exit cones and for off-design nozzles, it may be necessary to allow for this effect.

This is another instance of the remarkable insensitivity of rocket nozzle performance to flow disturbances that are of great importance in wind tunnel nozzle design.

Jet detachment or flow separation. It is not always possible to provide a nozzle with the optimum area ratio for each condition of ambient pressure. In particular, the exhaust nozzle of a rocket airplane that must fly at altitudes from sea level to perhaps 60,000 feet, or of a guided missile that starts at sea level and reaches 100,000 at burnout, generally operates in a very much overexpanded condition at the lower altitudes. The question arises whether the flow in the divergent section remains isentropic.

The possibility of a stationary shock wave perpendicular to the flow axis has been recognized for a long time, and the early literature on rocket performance indicates that this is to be expected. However, although such normal shocks have been observed in supersonic flow channels with small angles of divergence, normal shocks have not been found at all in rocket nozzles. Instead, it is found that when the back pressure (pressure of the surrounding atmosphere) exceeds the isentropic exit pressure by a large enough amount, the flow separates symmetrically from the nozzle wall, and a complicated pattern of oblique shock waves appears in the stream beyond the point of separation (Plate G,3).

The fundamental fluid mechanics of this type of separation phenomenon is discussed in IV,B. In brief, the region of importance is the place near the wall where the shock wave penetrates the boundary layer. Here the sharp adverse pressure gradient connected with the shock thickens the boundary layer and leads to deflection of the flow, while the pressure rise across the shock is supported by the momentum transfer from the

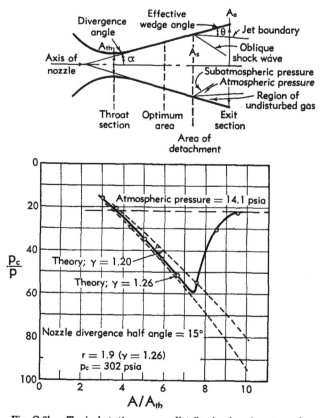

Fig. G,3b.   Typical static pressure distribution in exhaust nozzle.

supersonic stream to the "stagnant" fluid in the flow corner. Consequently the strength of the shock and the separation location depend on the state of the boundary layer (laminar or turbulent) and particularly on its development along the nozzle wall prior to separation. For the purpose of the reader interested in rockets, the subject can be presented most usefully in terms of the experimental results and the empirical correlations.

The static pressure distribution in the nozzle is shown in Fig. G,3b. The pressure follows the isentropic expansion curve until the separation point is reached. At this point, the pressure rises steeply to almost the level of the ambient pressure at the exit, indicating a rather clean break-away of the flow. From the standpoint of thrust, the nozzle may well be cut off at the separation station, since the internal and external pressures are very nearly in balance beyond this point. Experimentally, it makes little difference if a longer nozzle is used, once separation has occurred. The extra length of nozzle does not affect the location of separation and has no appreciable effect on the thrust. (There is a small reduction in thrust due to the extra length of cone, because the internal pressure is slightly less than the external pressure, but this is usually neglected in performance estimates.)

It is obvious that the complicated flow structure downstream of separation is of no consequence in the computation of thrust. The thrust of a rocket motor with separation in the nozzle is calculated on the assumption that the nozzle area ratio is not that corresponding to the actual exit but rather that of the separation station. Therefore it is necessary to be able to predict the location of separation.

In tests with nitric acid-aniline rocket motors employing nozzles with half angles α of 10°, 15°, 20°, and 30°, with actual area ratios ϵ of 10 and 20, and with pressure ratios p_o/p_∞ from 11 to 25, it was found that

$$\frac{p_s}{p_\infty} \cong 0.36 \pm 0.02 \qquad (3\text{-}5)$$

Almost identical results were obtained at all mixture ratios, and in fact, the results obtained with the nitrogen flow channel previously mentioned were also nearly the same. Consequently, as a general rule for estimating the performance of overexpanded nozzles in the absence of specific test data, the relation (Eq. 3-5) may be used.

Heat loss from the combustion gas. The transfer of heat from the combustion gas to the relatively cold motor walls, by convection and by radiation, affects the performance of the motor. Two cases of heat trans-fer are of practical interest: (a) the heat transferred is completely removed from the thermodynamic process, as in a water-cooled motor; and (b) the heat is reintroduced into the process as in a motor cooled by the propellants (regenerative cooling). These two processes, together with the adiabatic one, are shown in the h, s diagrams of Fig. G,3c.

In diagram 1, the case of no heat loss, the working fluid (product gas) is injected into the chamber cold at A, is heated adiabatically at constant pressure by combustion to point B, and is expanded isentropically to ambient pressure at C. The enthalpy drop from B to C is transformed to the kinetic energy of the jet (see Eq. 2-10). In diagram 2, the case of heat lost to the cooling water, it is assumed in order to simplify the calcu-lations that the heat removal is concentrated at two stations, the hot end of the chamber and the throat of the exhaust nozzle. (This is reasonable on the basis of experimental heat transfer distributions.) The process is then as follows: adiabatic combustion from A to B, constant pressure heat loss from B to D, isentropic expansion from D to E, constant pressure heat loss from E to F, and isentropic expansion from F to G. The regenerative process, diagram 3, traces the following path: constant pressure heating from A to J, adiabatic combustion from J to K, heat loss from K to L, expansion from L to M, heat loss from M to N, and expansion from N to O.

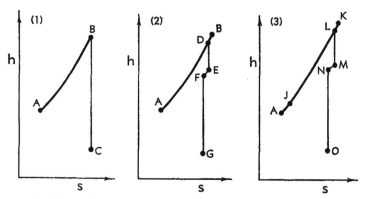

Fig. G,3c. Thermodynamic processes for (1) adiabatic, (2) water-cooled, and (3) regeneratively cooled rocket motors.

The exhaust velocities for the three diagrams are:

$$\left.\begin{aligned} \tfrac{1}{2}V_{e(1)}^2 &= h_B - h_C \\ \tfrac{1}{2}V_{e(2)}^2 &= (h_D - h_E) + (h_F - h_G) \\ \tfrac{1}{2}V_{e(3)}^2 &= (h_L - h_M) + (h_N - h_O) \end{aligned}\right\} \tag{3-6}$$

The heat of combustion:
$$\Delta h_C = h_B - h_A = h_K - h_J \tag{3-7}$$

In cases 2 and 3, the heat transferred per unit mass is given by:

$$\left.\begin{aligned} \Delta q_2 &= (h_B - h_D) + (h_E - h_F) \\ \Delta q_3 &= (h_K - h_L) + (h_M - h_N) = (h_J - h_A) \end{aligned}\right\} \tag{3-8}$$

For a perfect-gas working fluid with constant specific heat, the ratio of the exhaust velocity of the cooled motor to that of the ideal motor can be given explicit form, in each case, in terms of the heat of combustion and the pressure ratio. In general, to calculate the correction with precision, it is necessary to refer to actual h,s computations on diagrams that incorporate the temperature-dependent specific heats and the effects of dissociation. It is immediately evident from the diagrams that the water-cooled motor produces a lower specific impulse than the ideal one. In practice, the approximate formula: $V_{e(2)} \cong V_{e(1)}[(\Delta h_c - \Delta q_2)/\Delta h_c]^{\frac{1}{2}}$ is applied to relate the two results. For small test motors, this can amount to as much as 5 per cent. It is also evident that the specific impulse of a regeneratively cooled motor is never less than that of the ideal one, and can *exceed* the ideal performance if the heat transfer $h_M - h_N$ is large in comparison with $h_K - h_L$. In practice, this effect is too small to be measured.

Suspended liquid or solid particles in the exhaust jet. Condensed phases can appear in the exhaust jet if elements are present in the reactants that form refractory products. For example, fuel compounds based on the light metals Li, Be, Al, and Mg produce oxides Li_2O, BeO, Al_2O_3, and MgO when reacted with oxygen. These oxides have normal vaporization temperatures of 2250, 3900, >1700, and >2800°C, respectively, and therefore these vapors would condense to droplets or solid particles whenever the temperature drops below these levels, either in the combustion or in the expansion process. Similarly, carbon (sublimation point >4200°C) which can form in large amounts in the decomposition of acetylenic monopropellants, and KCl (sublimation point, 1500°C) which forms in solid propellant combustion where potassium perchlorate is used as the oxidizer, present the problem of flow with condensed phases.

The occurrence of condensation leads to a reduction in specific impulse for two reasons. The portion of the working fluid that condenses cannot perform any expansion work and therefore cannot contribute to the acceleration of the jet; also, the heat in the condensed phase is partly ejected with the jet and not transformed to kinetic energy, because of the low rate of heat transfer from the hot particles to the surrounding gas. The rate of heat transfer depends on the particle size, which in turn depends on whether condensation occurs in the chamber or at some position in the nozzle. Average particle sizes have been observed in the range from 10μ to 50μ.

To analyze the problem in the most exact way, it is necessary to know at what place in the expansion process nucleation occurs, the rate of growth of the condensed nuclei, the rate of heat transfer from the hot particles to the cooler gas stream, and the velocity lag between the suspended particles and the gas stream. For practical purposes, it is usually adequate to simplify the analysis by assuming that the velocity lag is zero. Concerning heat transfer, two extreme assumptions are possible: (1) the temperature of the condensed phase remains the same throughout the expansion process as that of the gas where condensation took place, that is, zero heat transfer; and (2) the condensed phase temperature follows exactly the temperature history of the gas, that is, perfect heat exchange. The first assumption would apply to very large particles; the second assumption would apply to very small particles.

The equation of state of the mixture of gas and suspended particles, and the density of the mixture ρ, are definable in terms of the mean molecular weight $\mathfrak{M}_g$ of the gas, the molecular weight of the solid $\mathfrak{M}_s$, the mean molecular weight of the mixture $\overline{\mathfrak{M}}$, and the respective numbers of moles of gas and solid, n_g and n_s, in a standard molar volume $\mathcal{V}_0$.

$$p = \frac{n_g}{\mathcal{V}_0} RT_g = p_g \\ \rho = \frac{n_s\mathfrak{M}_s + n_g\mathfrak{M}_g}{\mathcal{V}_0} \left.\right\} \tag{3-9}$$

For adiabatic, frictionless flow, $dh = (1/\rho)dp$. For the mixture,

$$\frac{n_g C_{p_g}dT_g + n_s C_{p_s}dT_s}{n_g\mathfrak{M}_g + n_s\mathfrak{M}_s} = \frac{n_g RT_g}{n_g\mathfrak{M}_g + n_s\mathfrak{M}_s}\frac{dp_g}{p_g} \tag{3-10}$$

Introducing mole fractions X_g and X_s, this equation becomes

$$X_g C_{p_g}\frac{dT_g}{T_g} + X_s C_{p_s}\frac{dT_s}{T_s} = X_g R\frac{dp_g}{p_g} \tag{3-11}$$

For adiabatic flow, $VdV + dh = 0$. For the mixture, with zero velocity lag,

$$\tfrac{1}{2}(X_g\mathfrak{M}_g + X_s\mathfrak{M}_s)VdV + X_g C_{p_g}dT_g + X_s C_{p_s}dT_s = 0 \tag{3-12}$$

Eq. 3-11 and 3-12 are applicable upstream of the condensation location, if X_g is set equal to unity, and X_s equal to zero. Downstream of this location, these equations can be applied if it is assumed that the mole fractions X_g and X_s are fixed. This is a reasonable assumption since most of the condensation takes place in a very short region.

A satisfactory solution of the problem is to obtain V_e in terms of the pressure ratio p_e/p_c. This can be obtained by integrating Eq. 3-11 and 3-12 if the ratio dT_s/dT_g can be specified. For case 1 mentioned above, the ratio is zero; for case 2, the ratio is unity; and other cases can be treated. The most extreme case is to assume that dT_s/dT_g is zero and that condensation occurs in the combustion chamber, so that the particles are at temperature T_c at the exit. This leads to:

$$V_e = \left\{ \frac{2X_g C_{p_g}T_c}{\overline{\mathfrak{M}}}\left[1 - \left(\frac{p_e}{p_c}\right)^{\frac{\gamma-1}{\gamma}}\right]\right\}^{\frac{1}{2}} \tag{3-13}$$

$$\overline{\mathfrak{M}} = X_g\mathfrak{M}_g + X_s\mathfrak{M}_s$$

$$\gamma = C_{p_g}/C_{v_g}$$

For a typical propellant gas composition, based on hydrocarbon combustion, the condensation of 20 per cent of the mass in the combustion chamber leads to about 10 per cent loss in I_{sp}, by formula (Eq. 3-13). The effect can therefore be of considerable importance for some propellant systems.

Pressure drop in the combustor due to heat release. The analysis of the ideal rocket motor is based on the postulate that combustion takes place at constant pressure. In practice, the release of heat in the gas stream in the combustor is accompanied by a decrease of both static and stagnation pressures. The decrease is dependent on the Mach number at the hot end of the combustion chamber, and becomes larger as the ratio of combustor cross-sectional area A_c to throat area A_{th} approaches unity. The loss in stagnation pressure represents an undesirable increase in entropy, a decrease in the effective pressure ratio for expansion, and hence a loss in specific impulse. The following analysis deals with these effects.

As before, the subscripts i, c, th, and e refer to states at the injector end of the combustor, the hot end of the combustor, the nozzle throat, and the nozzle exit. Except for the nonzero velocity in the combustor, the conditions of the analysis are the same as for the ideal rocket motor.

Since the expansion in the nozzle is isentropic (heat release completed in the chamber) the Mach number at c is obtainable from Eq. 2-10:

$$\frac{A_c}{A_{th}} = \frac{1}{M_c} \left(\frac{1 + \dfrac{\gamma - 1}{2} M_c^2}{\dfrac{\gamma + 1}{2}} \right)^{\frac{\gamma+1}{2(\gamma-1)}} \tag{3-14}$$

The integrated momentum equation, for frictionless flow in a constant area duct, provides an expression for the decrease in static pressure in the chamber. It is assumed that $M_i \cong 0$.

$$p_i = p_c(1 + \gamma M_c^2) \tag{3-15}$$

The simplest way to treat the performance for the case of nonzero M_c is first to calculate the equivalent plenum chamber pressure and temperature corresponding to the conditions at c, and then to calculate the flow properties in the nozzle on the assumption that the gas flow originates from this plenum chamber. The temperature T_c^0 and pressure p_c^0 of the equivalent plenum chamber are really the stagnation temperature and pressure at station c. (See III,A for discussion of stagnation quantities.)

$$p_c^0 = p_c \left(1 + \frac{\gamma - 1}{2} M_c^2 \right)^{\frac{\gamma}{\gamma-1}} \tag{3-16}$$

$$T_c^0 = T_c \left(1 + \frac{\gamma - 1}{2} M_c^2 \right) \tag{3-17}$$

$$\Delta h_c = c_p(T_c^0 - T_i) \tag{3-18}$$

Eq. 3-18 is simply a statement of the conservation of energy, with the kinetic energy term $\frac{1}{2}V_c^2$ absorbed in the stagnation enthalpy $c_p T_c^0$.

Therefore, T_c^0 is the same as T_c in the ideal rocket motor of Eq. 2-7, that is, the so-called adiabatic flame temperature, or simply the chamber temperature. The static temperature T_c is less than the adiabatic flame temperature, by as much as 10 or 15 per cent for a throatless motor, but this is of no consequence for the exhaust velocity. Only the stagnation temperature matters, and this depends only on the thermochemical heat of reaction Δh_c.

Fig. G,3d compares the conditions of the equivalent plenum chamber with those of the actual chamber by presenting the ratios p_c/p_i, p_c^0/p_i,

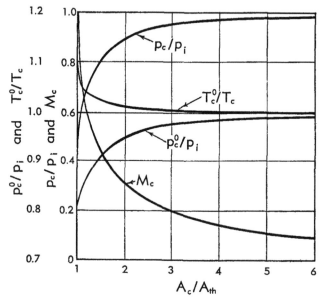

Fig. G,3d. Comparison of conditions in equivalent plenum chamber with those in actual combustion chamber.

M_c, and T_c/T_c^0 as functions of A_c/A_{th}. This figure is useful as a guide in carrying out the following computational procedure recommended for the performance analysis of a rocket motor.

In general, the particular propellant combination, the feed pressure, the nozzle throat area, the external pressure, and the chamber dimensions are specified in advance. The problem is to compute the effective exhaust velocity c, taking into account the effect of the pressure drop in the combustion chamber.

1. From the specified A_c/A_{th} and γ, compute M_c from Eq. 3-14.
2. With this M_c and the specified p_i, calculate p_c and then p_c^0 from Eq. 3-15 and 3-16.
3. From the specified ϵ of the nozzle, and the effective pressure ratio

p_c^0/p_∞, determine C_F^0 for the equivalent plenum chamber, from Fig. G,2d and G,2e.

4. The thrust is then calculated as follows:

$$F = C_F^0 p_c^0 A_{th} \tag{3-19}$$

5. The characteristic velocity c^* depends on the stagnation temperature, which is exactly equal to the adiabatic flame temperature. Therefore, $(c^*)^0$ is known, either by thermochemical calculation or by experiments with plenum-type combustors.

6. The mass flow is then calculated in terms of $(c^*)^0$:

$$\dot{m} = \frac{p_c^0 A_{th}}{(c^*)^0} \tag{3-20}$$

7. Finally, the effective exhaust velocity is calculated:

$$c = C_F^0 (c^*)^0 \tag{3-21}$$

In rocket testing, the chamber pressure is frequently measured at the forward end of the chamber, through a pressure tap drilled into the injector face. Clearly, this is p_i and not p_c^0. Unless the measured p_i is converted by Eq. 3-15 and 3-16 to p_c^0, the reported value of c^* will be too high by as much as 25 per cent (Fig. G,3b) for a throatless motor, and the reported value of C_F (experimental)$/C_F$ (theoretical) will be low by a comparable error. These effects are significant for a rocket motor with a throat diameter equal to $\frac{1}{2}$ or more of the chamber diameter. If the combustor is not cylindrical, the preceding analysis must be modified. In particular, Eq. 3-15 holds only for cylindrical combustors; for other shapes, it is necessary to know the axial distribution of the heat release in order to integrate the momentum equation.

It may be noted that the same theory can be used to determine the axial distribution of heat release, i.e. the over-all kinetics of combustion. Thus, in a cylindrical chamber, the variation of M with axial position x is obtainable from the pressure distribution by means of Eq. 3-15. Then, from the equation of continuity (Eq. 3-22), the point-by-point static temperature can be computed, and from this, the stagnation temperature and the heat release.

$$\dot{m} = \frac{p A_c \mathfrak{M}}{[RT/\gamma M]^{\frac{1}{2}}} \tag{3-22}$$

G,4. Theoretical Specific Impulse Calculations. The calculation of specific impulse in the theory of the ideal rocket motor in Art. 2 is deficient in two very important respects: the combustion gas mixture is chemically reactive, and the specific heats vary strongly with temperature. Both these factors were ignored there. It is the purpose of this article to develop the complete theory, with both effects present. Inasmuch as the thermodynamics of combustion and the theory of expansion of reactive mixtures are covered in detail in II,A, B, and C, the theory is given here only in condensed form. The reader is referred to Vol. II for a more complete discussion.

The problem is the following one: for a given set of reactants (propellants) and a specified constant combustion pressure, calculate the adiabatic flame temperature and the composition of the product mixture, and then, for expansion to a specified exit pressure, calculate the specific impulse and the proper nozzle area ratio.

The analysis is carried out most conveniently in terms of the *total enthalpy H* of the reacting mixture and the *total entropy S*, both quantities being computed for a definite amount of the mixture.

$$H = \sum_i n_i[(H_T^0 - H_{T_0}^0) + \Delta H_{iT_0}^0]_i \tag{4-1}$$

$$S = \sum_i n_i(S_{T,P}^0)_i = \sum_j n_j[S_{T,1}^0 - R \ln p_j]_j + \sum_k n_k(S_{T,1}^0)_k \tag{4-2}$$

where $\Delta H_{iT_0}^0$ = molar enthalpy of formation of component i at reference temperature T_0 (usually 0°K or 298°K) in the ideal state.

$H_T^0 - H_{T_0}^0$ = molar thermal enthalpy or heat content at temperature T with respect to temperature T_0, also in the ideal state. Equivalent to

$$\int_{T_0}^T C_p^0 dT + \sum \Delta H_T^0$$

The latter sum refers to all the heats of transition between T_0 and T.

$S_{T,P}^0$ = absolute molar entropy (exclusive of nuclear spin) of component i at temperature T and pressure p, in the ideal state.

$S_{T,1}^0$ = value of $S_{T,P}$ at the standard pressure of one atmosphere.

p_j = partial pressure (in atmospheres) of gaseous component j, considered to be an ideal gas. $[p_j = \rho_j RT/\mathfrak{M}_j]$

n_j = number of moles of gaseous component j in the specified mixture.

n_k = number of moles of condensed component k (liquid or solid) in the mixture.

n_i = number of moles of any component (gas, liquid, or solid); $\sum n_i = \sum n_j + \sum n_k$.

The following physical conditions are implicit in Eq. 4-1 and 4-2. Each substance is in its ideal state corresponding to the particular temperature and partial pressure that prevails. This is indicated by the superscript 0. Thus, each gaseous component obeys the perfect gas law, and in particular, the specific heats are dependent only on temperature and not on pressure. For the condensed phases, the specific heat also depends only on the temperature. In ideal mixtures, there is no energy of mixing, so that the total enthalpy is simply the sum of the component enthalpies, as given by Eq. 4-1.

Similarly, the total entropy is simply the sum of the component entropies. The gaseous components, being perfect gases, contribute logarithmic pressure terms, as in Eq. 4-2. The liquid and solid components in their ideal states have entropies that are independent of the pressure, and therefore the terms appear in a separate sum in Eq. 4-2.

For the hot combustion products encountered in rockets, the assumption of the ideal state and ideal mixing is quite accurate up to pressures of 200 atmospheres. Above that, the gas density is 10 per cent or more of the average critical point density of the gas mixture, and the intermolecular forces begin to affect seriously the equations of state and the enthalpies. Therefore, for pressures of 200 atmospheres or more, the analysis described here needs modification. Details are given in II,A. For the condensed products, the ideal state assumption is generally good to even higher pressures. On the reactant side, the liquid propellants may be treated as ideal substances with entropy and enthalpy both independent of pressure. This is acceptable because the pressure effect on the enthalpy is usually quite small, while the effect on the entropy does not matter in the analysis.

The notations and units employed in this article for the treatment of combustion are those normally familiar to physical chemists rather than to mechanical engineers. The reason is simply one of convenience: the standard published collections of thermochemical data make use of the c.g.s. system of units, the Kelvin temperature scale, molar quantities, and the same symbols as those in Eq. 4-1 and 4-2. Another important difference from the mechanical engineering literature pertains to the zeros of the enthalpy and entropy scales. For example, in combustion engineering, there exists the convention of assigning zero to the enthalpy of formation of CO_2, H_2O, O_2, and N_2. In this article, the physical chemists' assignment of zero to all the elements in their standard states is used. Thus, in the combustion engineer's system, H_2 gas has a positive heat of formation, but zero in the system in use here. Finally, when in the treatment of the expansion process the mass flow and the jet velocity are to be calculated, the fps system of units is employed in this article in order to conform to current rocket practice.

As a first step, the combustion reaction is written down with all the products that are likely to be present in appreciable amounts. For example:

$$n_{H_2}H_2 + n_{O_2}O_2 \rightarrow n_{H_2O}H_2O + n_{OH}OH + n_{H_2}H_2 + n_H H + n_{O_2}O_2 + n_O O$$
(reactants)   (products)

The weight ratio r of oxidizer to fuel is $n_{O_2}\mathfrak{M}_{O_2}/n_{H_2}\mathfrak{M}_{H_2}$.

The energy equation is

$$H_c \text{ (products)} = H_i \text{ (reactants)} - \Delta Q_{\text{loss}}$$

$$\sum_{\text{prod}} n_k[(H_{T_c}^0 - H_{T_0}^0) + \Delta H_{iT_0}^0]_k = \sum_{\text{react}} n_j[(H_{T_i}^0 - H_{T_0}^0) + \Delta H_{iT_0}^0]_j - \Delta Q_{\text{loss}}$$

$$(4\text{-}3)$$

The flame temperature T_c and the composition n_k's are unknown. The injection temperatures T_{ij} are specified, and so the entire right member of Eq. 4-3 can be computed. ΔQ_{loss} is a measured heat loss, if any.

Additional equations that make possible the determination of the n_k's and T_c are those governing the chemical equilibria at a specified pressure p_c:

(a) Mass conservation, one for each element in the system (linear equations). For example: $2n_{H_2} = 2n_{H_2O} + n_{OH} + 2n_{H_2} + n_H$

(b) Chemical equilibrium, one for each *independent* reaction (nonlinear equations). For example: $\dfrac{n_H n_{OH}}{n_{H_2O}} = K_p \dfrac{\sum n \text{ (products)}}{P_c}$

All together, there are enough equations to solve for the unknown composition and flame temperature. Because the chemical equilibria are nonlinear and because the energy equation is not explicit in T_c, numerical or graphical methods of solution are necessary. Specific procedures are discussed in II,A and C.

The next step is the analysis of the expansion process. If the expansion is adiabatic and nondissipative, then it is isentropic:

$$S_e \text{ (exhaust products)} = S_c \text{ (combustor products)}$$

Concerning the physical nature of the expansion process, several alternative assumptions are possible. First, if condensed phases are produced either in the combustor or in the nozzle, special analysis is required (see Art. 3). If there is no condensation, then it is necessary to decide whether chemical equilibrium is maintained during the entire expansion. If not, a satisfactory approximation would be to postulate full chemical equilibrium down to a certain temperature level and then frozen composition below that level. A more exact analysis would involve the explicit introduction of the chemical kinetic relationships in place of the chemical equilibrium equations. This problem is discussed in III,B. Finally, it is necessary to decide whether the internal degrees of freedom of the molecules can adjust their energies to the rapid expansion, that is, whether there is an appreciable vibrational energy lag. This is also discussed in III,B.

Both the chemical and vibrational lags are dependent on the expansion time. For motors of less than 50-lb thrust, nozzles less than 2 inches long, expansion times of 10^{-4} sec or less, the chemical lags might be appreciable, but for motors of 1000-lb thrust or more, it is likely that these effects are unimportant. The vibrational lags do not seem to occur at all, probably because of the high temperature level and the consequent rapid rates of adjustment. In general, the chemical lag is difficult to detect by its effect on specific impulse because the difference in I_{sp} is rarely more than a few per cent, even for very hot systems with considerable dissociation.

The most effective ways to detect a lag in radical recombination would be by the exit temperature or by the pressure distribution in the exhaust nozzle. However, both types of experiments have been tried and both leave much to be desired in the interpretation of the results. It seems the best practice, therefore, to compute both the constant composition performance and the equilibrium composition performance, and to regard them as lower and upper limits, respectively. Another possible method is to determine experimentally the area ratio that expands the gas to an exit pressure of exactly one atmosphere, this condition being observable from the exhaust shock wave pattern.

Expansion with complete chemical and vibrational equilibrium, and no condensation of solid or liquid substances, is governed by the following entropy equation and a set of equations governing the chemical equilibrium:

$$\sum_{\text{exhaust}} n_l \left[S^0_{T_e,1} - R \ln \left(\frac{p_e n_l}{\sum n_l} \right) \right]_l = \sum_{\text{combustor}} n_k \left[S^0_{T_c,1} - R \ln \left(\frac{p_c n_k}{\sum n_k} \right) \right]_k$$

$$(4\text{-}4)$$

The exhaust temperature T_e and the exhaust composition n_l's are unknown. The entire right member is known from the solution of the first part of the problem. The additional equations are, as before,

(a) Mass conservation, one for each element in the system (linear equations).
(b) Chemical equilibrium, one for each *independent* reaction (nonlinear equations).

As before, numerical or graphical methods are required to solve this system of equations for the n_l's and T_e (see III,B for details).

The exhaust jet velocity V_e is obtainable from the energy relation for the expansion, after evaluating H_e from the now-known exhaust composition and temperature.

$$\tfrac{1}{2} V_e^2 \sum_{\text{react}} n_j \mathfrak{M}_j = H_c \text{ (combustor)} - H_e \text{ (exit)} \qquad (4\text{-}5)$$

The mean molecular weight, the gas density, and the area per unit mass flow, all at the exit, are given by

$$\overline{\mathfrak{M}}_e = \frac{\sum n_i \mathfrak{M}_i}{\sum n_i} \tag{4-6}$$

$$\rho_e = \frac{\overline{\mathfrak{M}}_e p'_e}{RT_e} \tag{4-7}$$

$$\frac{A_e}{\dot{m}} = \frac{1}{\rho_e V_e} \tag{4-8}$$

To determine the proper nozzle area ratio, the method just outlined can be utilized. A series of pressures p is constructed, ranging from p_c down to p_e, and then Eq. 4-4, 4-5, 4-6, 4-7, and 4-8 are solved for each pressure. A curve is then plotted of $A/\dot{m}$ against p. The minimum of this curve corresponds to the throat of the de Laval nozzle and is designated as $A_{th}/\dot{m}$. The ratio of $A_e/\dot{m}$ determined in Eq. 4-8 to $A_{th}/\dot{m}$ constitutes the desired nozzle area ratio ϵ_{opt}.

This procedure produces, of course, more information than merely ϵ_{opt}. The theoretical value of c^* is obtained as the product $p_c(A_{th}/\dot{m})$, and the theoretical value of C_F is obtained as V_e/c^*. Also, the axial distribution of pressure along the nozzle is given by the curve just plotted.

The general scheme of the performance analysis can be summarized by tracing the process on a suitable Mollier diagram. To construct this diagram, a definite atomic composition is specified, that is, a definite number of gram-atoms of C, H, O, N, etc., respectively. These atoms are assumed to form components of the product mixture that are in chemical equilibrium at every point of the diagram. In principle, lines of constant pressure can be mapped in the following way: Specify a total enthalpy H as defined in Eq. 4-1 and a total pressure p, then solve for the total entropy S. This involves, of course, solving for the equilibrium composition and the temperature. Because the equilibrium constants depend on the temperature in a complicated way, it is easier to locate the isobars by starting with a given T and a given p, and then solving for H and S. It is to be noted that this Mollier diagram differs from that shown in Fig. G,3a for the analysis of the ideal rocket motor. In the latter diagram, the enthalpy included only the thermal enthalpy and not the chemical enthalpy, that is, the gas mixture was assumed to be nonreactive and the mixture composition was the same everywhere on the diagram. In Art. 3, the enthalpy of the propellant gas was raised from h_A to h_B by the heat of combustion (see Eq. 3-7); in the present analysis, because the chemical enthalpy is included, the total enthalpy of the reactants equals the total enthalpy of the products, except for heat transfer.

Once the Mollier diagram is constructed, the analysis proceeds as follows. By means of Eq. 4-1, the total enthalpy of the propellant combination is computed, and from this is subtracted the heat loss, if any. The resulting enthalpy is marked on the H scale. A horizontal line from this starting point (i) to the intersection (c) with the proper isobar p_c can be considered to represent the process of combustion. Then, a vertical line from state c to the intersection (e) with the proper isobar p_e represents the isentropic expansion. The enthalpy drop from c to e is the enthalpy converted to jet kinetic energy, according to Eq. 4-5. It is clear that the same Mollier diagram can be used for any propellant combination having the proper atomic constitution, the only difference being in the location of the injection state (i) on the H scale. It should be noted that, as a practical matter, it is not necessary to construct the entire Mollier diagram unless it is to be used ultimately for a large number of performance analyses for widely varying propellant systems and combustion pressures. Otherwise, it is sufficient to compute merely the states c and e, and such other specific states as may be of interest.

It is sometimes convenient to report the results of a performance analysis in terms not only of V_e, but also of an effective value of the specific heat ratio $\bar{\gamma}$ so as to fit a formula like Eq. 2-24:

$$c^* = \left[\frac{1}{\bar{\gamma}} \left(\frac{\bar{\gamma} + 1}{2} \right)^{\frac{\bar{\gamma}+1}{\bar{\gamma}-1}} \frac{RT_c}{\mathfrak{M}} \right]^{\frac{1}{2}} \qquad (4\text{-}9)$$

A definition of $\bar{\gamma}$ that meets this requirement is the following:

$$\bar{\gamma} = \frac{\bar{C}_p}{\bar{C}_p - R}, \quad \bar{C}_p = \frac{(H_c - H_e)}{(T_c - T_e)(\overline{\sum n_i})}, \quad \overline{\sum n_i} = \frac{1}{2} \left(\sum_c n_i + \sum_e n_i \right) \qquad (4\text{-}10)$$

The sum of the number of moles in the specified amount of gas in the combustion chamber is $\sum_c n_i$. Due to radical recombination during expansion, the similar sum $\sum_e n_i$ at the exit is somewhat less. $\overline{\sum n_i}$ is the arithmetic mean of the two and $\overline{\mathfrak{M}}$ is based on this sum. It is obvious that $\bar{C}_p$ includes the heat of recombination and is therefore considerably larger than the specific heat for the case of no reaction. Correspondingly, the value of $\bar{\gamma}$ is lower than the no-reaction γ. The exit temperature T_e is consistent with this value of $\bar{\gamma}$:

$$\frac{T_e}{T_c} = \left(\frac{p_e}{p_c} \right)^{\frac{(\bar{\gamma}-1)}{\bar{\gamma}}} \qquad (4\text{-}11)$$

Another way to compute the state of the gas mixture at e is to make use of Eq. 4-10 and 4-11. The procedure starts by taking a trial value of $\bar{\gamma}$, and then computing in succession T_e, n_i at e, $\overline{\sum n_i}$, $(H_c - H_e)$, $\bar{C}_p$, and again $\bar{\gamma}$. If the last value of $\bar{\gamma}$ does not agree with the first value, a new trial value is selected until consistency is obtained. In much of the literature, a performance analysis is reported not only in terms of I_{sp} and T_c, but with $\bar{\gamma}$ and $\overline{\mathfrak{M}}$ as well.

It was pointed out above that the alternate assumption of constant composition expansion is also possible. The analysis of this case is greatly simplified because the composition of the end state is known. The same equation (Eq. 4-4) is used to evaluate the temperature and the enthalpy at the exit, or Eq. 4-10 and 4-11 can be used for this purpose. In either case, the general result is that the exit temperature T_e for a constant-composition expansion is lower than that for an equilibrium-composition expansion, particularly if there is appreciable dissociation in the state at T_c. However, the enthalpy difference is always slightly *less* for the constant-composition expansion, and therefore, by Eq. 4-5, the constant-composition case produces a slightly lower specific impulse.

The difference is usually only a few per cent. As a qualitative explanation, the larger I_{sp} of the equilibrium composition case is due to the heat released during the expansion by the recombination of free atoms and radicals, but although the recombination heat may be large, the effect on I_{sp} is weakened by the low pressure level at which it is released, that is, the thermodynamic conversion of the recombination heat is not as favorable as that of the main heat of combustion in the chamber. Of course, for a rocket operating with very low exhaust pressure, such as the upper stage of a satellite launcher, the large pressure ratio permits a favorable conversion to kinetic energy of the recombination heat released in the nozzle. In this case, the difference in I_{sp} may be significant.

Table G,4a. Calculated performance characteristics of several monopropellants and bipropellant combinations.

Propellant	I_{sp}	T_c	γ	r	$\overline{\mathfrak{M}}$
Monopropellants: Ethylene oxide	160	1860	1.17	—	44
Hydrazine	170	1125	1.37	—	32.1
Nitromethane	218	3950	1.25	—	61
Bipropellants: Liquid oxygen + 75% ethyl alcohol + 25% water	227	4875	1.19	1.38	23.8
Liquid oxygen + hydrazine	265	5200	1.23	0.68	16.6
Liquid oxygen + liquid hydrogen	346	4150	1.28	3.30	8.6
Liquid oxygen + gasoline	243	5450	1.22	2.50	22.7
RFNA + 15% NO_2 + aniline	222	5050	1.23	3.00	24.5
RFNA + 13% NO_2 ethyl alcohol	218	4750	1.21	2.80	24.4

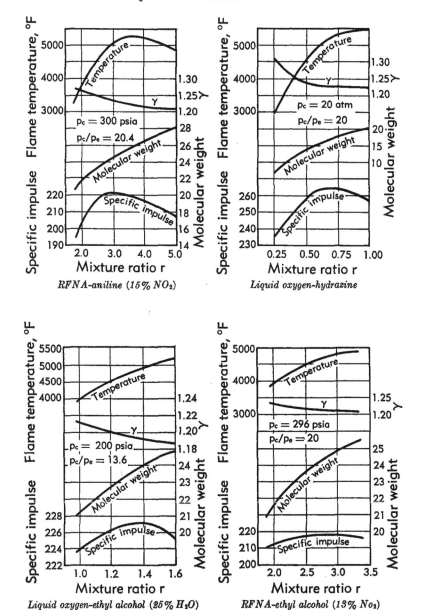

Fig. G,4. Variation of performance characteristics with mixture ratio for several bipropellant combinations. After Sutton, G.P., *op. cit.*

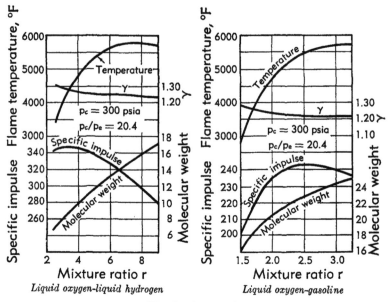

Fig. G,4 (continued)

This concludes the general theory of specific impulse calculations. The remainder of this article is devoted to a presentation of some typical results and a discussion of the influence of certain factors on the specific impulse and the flame temperature.

Calculated performance characteristics of a series of monopropellants and bipropellant combinations are listed in Table G,4a. The significant characteristics are: I_{sp}, T_c, $\bar{\gamma}$, r, and $\overline{\mathfrak{M}}$. Unless otherwise noted, the reported

Table G,4b. Equilibrium gas composition in the combustion chamber of several monopropellants.

Propellant	Ethylene oxide C_2H_4O	Hydrazine N_2H_4	Nitromethane CH_3NO_2
Theoretical decomposition products in mole fractions	0.481 CO	0.656 H_2	0.057 CO_2
	0.404 CH_4	0.331 N_2	0.277 CO
	0.077 H_2	0.013 NH_3	0.277 H_2O
	0.038 C_2H_2		0.223 H_2
			0.166 N_2

values correspond to expansion with complete equilibrium from 300 lb/in.^{2} abs chamber pressure to 14.7 lb/in.^{2} abs exhaust pressure, at the optimum mixture ratio. The stoichiometric ratio is shown in the table for comparison.

Table G,4b lists the equilibrium gas composition in the combustion chamber for the monopropellants C_2H_4O, CH_3NO_2, and N_2H_4.

Curves are presented in Fig. G,4 of the variation of I_{sp}, T_c, $\bar{\gamma}$, and $\overline{\mathfrak{M}}$ with mixture ratio r for the bipropellant combinations listed in Table G,4a.

Table G,4c. Equilibrium gas composition in the combustion chamber of several bipropellant combinations at mixture ratios for optimum impulse.

Propellant	Gasoline-oxygen 300 psi	Hydrogen-oxygen 300 psi	RFNA-aniline 300 psi	Fluorine-ammonia 500 psi
	CO 0.322	H_2O 0.312	CO_2 0.159	HF 0.676
	CO_2 0.137	H_2 0.687	CO 0.251	H_2 0.041
Products of reaction in mole fractions	H_2 0.105	H 0.001	H_2O 0.283	N_2 0.137
	H_2O 0.330		H_2 0.063	H 0.085
	H 0.028		N_2 0.192	F 0.054
	OH 0.061		H 0.011	N 0.007
	O_2 0.008		OH 0.032	
	O 0.006			

Finally, equilibrium gas compositions are given in Table G,4c for oxygen-gasoline, oxygen-hydrogen, RFNA-aniline, and fluorine ammonia at mixture ratio for optimum impulse.

Several conclusions of general applicability can be deduced from an examination of these performance characteristics and the preceding discussion of the combustion and expansion processes:

1. The peak flame temperature occurs close to the stoichiometric mixture ratio in C-H-O-N systems, although some deviations can be observed. Thus, because of the relatively strong stability of nitric acid compared with aniline, this system shows a peak temperature on the fuel-rich side of stoichiometric, whereas no appreciable deviation is observed in oxygen-gasoline. This is significant for the cooling problem.

2. The peak specific impulse in C-H-O-N systems always occurs far to the fuel-rich side of stoichiometric. Although the temperature is less than the stoichiometric value, the mean molecular weight $\overline{\mathfrak{M}}$ is even less proportionately, due to the higher concentrations of H_2 and CO on the rich side instead of CO_2 and O_2 on the lean side. The strong effect

of a reduction in $\overline{\mathfrak{M}}$ in raising the specific impulse despite a reduction in temperature is seen in the decomposition of hydrazine mono-propellant. Water addition to an alcohol-oxygen propellant to depress the temperature as an aid to cooling results in only a modest reduction of specific impulse because of the low molecular weight of H_2O and its dissociation products.

3. The pronounced effect of low $\overline{\mathfrak{M}}$ on specific impulse suggests the use of noncarbonaceous fuels such as NH_3, N_2H_4, and H_2 in order to avoid the formation of heavy CO and CO_2 molecules. The calculated performance data bear out this prediction.

4. The difference in specific impulse between equilibrium composition flow and constant composition flow is at most a few per cent, in the expansion from 300 lb/in.² abs to 14.7 lb/in.² abs, but exhausting to a lower pressure magnifies the difference because of the more favorable conversion of the recombination energy into kinetic energy. Raising the combustion pressure tends to produce a similar effect, but a large increase in pressure narrows the difference again because the dissociation is suppressed.

5. In highly dissociated mixtures, c^* increases strongly with increasing pressure if equilibrium flow is assumed. In nondissociated systems, c^* is practically insensitive to pressure. Therefore, the shape of the c^* versus r curve can change considerably with increasing pressure.

6. At 300 lb/in.² abs pressure, flame temperatures in the C-H-O-N system cannot get much above 3500°K because of the strong increase in CO_2 and H_2O dissociation. This is true regardless of the fuel or oxidizer employed. Higher temperatures are possible at higher pressures where dissociation is suppressed.

7. Since flame temperatures in the C-H-O-N system at 300 lb/in.² abs are limited, and since no reaction is known that will produce lighter products than the hydrogen-oxygen system, there exists an upper limit of $(T_c/\overline{\mathfrak{M}})$ and therefore of I_{sp} at 300 lb/in.² abs. This is about 340 lb sec/lb.

8. The limits on T_c and I_{sp} for C-H-O-N systems can perhaps be evaded by not using compositions that will form CO_2 or H_2O. Thus, the light metals, Li, Be, and B, have become interesting as high energy fuels. Unfortunately, some of the products are not gases, e.g. B_2O_3, and in their most convenient form as liquids they contain hydrogen, e.g. B_2H_6. Switching to fluorine as the oxidizer allows the use of these high energy fuels with very little dissociation. Nevertheless, no system is yet known that exceeds the performance of fuel-rich, hydrogen-fluorine (350 lb sec/lb).

G,5. Combustor Design Principles. With regard to the individual processes that take place in a rocket motor, the main components of the motor are the injector, the combustion chamber, the exhaust nozzle, the cooling system, the igniter, and the starting device. The problem of exhaust nozzle design is treated in Art. 2 and 3, and the cooling problem is discussed in Art. 6. The starting system is included as part of the total propellant control and feed system in Art. 7, but the combustion aspects are considered here. This article is devoted, therefore, mainly to the problems of injection, combustion, and ignition.

Injection. The injector usually has the form of a single plate or shell containing many orifices or slots through which the propellants flow into the combustion chamber. As a rule, to provide adequate residence time for complete burning of all particles of propellant, it is located at the forward end or head of the chamber. To prevent burnout, it is cooled regeneratively or otherwise protected by the methods described in Art. 6.

The functions of the injector are: (1) to introduce the propellants in the desired ratio into the combustion chamber and into the combustion space, (2) to cause the necessary degree of liquid mixing, or interpenetration plus atomization, or both, in order to achieve rapid and complete reaction, (3) to distribute the propellant spray uniformly in the provided combustion space to make the combustor as compact as possible, and (4) to provide the proper pressure drop to fix the flow ratio and to dampen flow disturbances.

Experience with many injectors leads to the following understanding of the combustion process, according to whether the bipropellant combination is self-igniting or nonself-igniting, or whether a monopropellant is to be used.

Considering the self-igniting combination of nitric acid and UDMH (unsymmetrical dimethyl hydrazine), the rapidity with which flame occurs upon contact, suggests that the first step should be liquid-phase mixing, that is, the generation of as much interfacial contact in the liquid phase as possible. This is borne out by the better efficiency of the injectors with multiple pairs of unlike impinging streams as compared with the atomizing-type spray injection. Of course, the complete reaction cannot take place entirely in the liquid phase: heat generation immediately vaporizes the liquids so that the largest part of the reaction must take place in the gas phase. The unlike impinging stream injector was the first to be used successfully in this country with nitric acid propellants.

A nonself-igniting propellant combination such as liquid oxygen plus kerosene (or JP-4) can burn only in the mixed vapor phase. Consequently, the best injector for this combination is one that vaporizes the liquids as rapidly as possible and disperses them without regard to liquid-phase mixing. Suitable for this purpose is the showerhead-type injector that subdivides the incoming propellant into a large number of very thin streams, or the swirl-nozzle type that atomizes the injected streams, as well as the impinging-stream type. Burning probably takes place in a diffusion flame surrounding each droplet, with the last stages of reaction occurring in the hot gas phase.

With a monopropellant such as nitromethane, mixing is obviously not important, but atomization and distribution of the spray are. To date, some success has been achieved with atomizing swirl injectors.

The importance of proper distribution of the spray in the combustion chamber was mentioned. Poor distribution may result in ineffective utilization of the chamber volume so that an oversize chamber may be needed to attain full efficiency. This may create not merely a size and weight problem, but more seriously, a cooling problem.

The pressure drop across the injector under operating conditions is an important design characteristic. A typical magnitude is about 100 lb/in.[2] A high pressure differential has the advantage of producing a high velocity stream with good mixing and atomization, but the disadvantage of imposing a higher feed pressure requirement on the pumping system. Too low a differential runs the danger of allowing the flow rates or the mixture ratio to wander from the design point, since the injector pressure drop is an essential part of the hydraulic balance of the feed system, as explained in Art. 7. Too low a differential may also invite combustion instability of the low frequency type. (See below.)

The pressure drop Δp can be calculated if the empirical pressure loss coefficient K is known. For the fuel side,

$$\Delta p_t = (p_t - p_o) = K_f \tfrac{1}{2} \rho_t V_t^2 \tag{5-1}$$

$$V_t = \frac{\dot{m}_t}{\rho_t A_t} \tag{5-2}$$

The stream velocity V_t is determined by the total mass rate of flow of fuel $\dot{m}_t$ divided by the sum of the fuel orifice areas A_t, on the acceptable assumption that there is no appreciable contraction of the stream as it emerges from the orifice. The pressure loss coefficient is always greater than unity since even the smoothest orifices cause some momentum loss. In the hydraulics literature, the discharge coefficient C_d is frequently used:

$$V_t = C_{d_t} \left(\frac{2\Delta p_t}{\rho_t} \right)^{\tfrac{1}{2}} \tag{5-3}$$

$$C_d = \frac{1}{\sqrt{K}} \tag{5-4}$$

Typical orifice sizes for most injectors range between $\tfrac{1}{16}$ to $\tfrac{1}{8}$ inch. Holes that are too small may easily become plugged during operation by foreign matter; holes that are too large produce a poorly atomized or mixed spray. The quality of the streams issuing from the orifices is always affected by the contour of the orifice approach and the shape of the connecting liquid manifolds.

A fixed area injector limits the possibility of controlling the thrust of a liquid propellant rocket motor. The quadratic pressure loss relation (Eq. 5-1) shows that a reduction of the flow rate by 50 per cent would reduce the differential to 25 lb/in.^{2}, if it were 100 lb/in.^{2} at rated flow. Trouble would be encountered at this low level with mixture ratio deviations and combustion instability. The solution is obviously a mechanically variable-area injector or a multiple injection system, and both have been used.

An important step in the development of an injector is the water test. By forcing water through the injector at about the rated flow, the discharge coefficient can be measured, and mixing atomization and distribution characteristics can be observed. Visual determination of the quality of the spray is aided by coloring the fuel and oxidizer water supplies with contrasting dyes, e.g. yellow and purple. It may be necessary to discharge into a simulated pressurized chamber in order to prevent misleading cavitation in the orifices at atmospheric pressure.

Another useful test method employed in injector development is the combustion chamber traverse with cooled temperature and composition probes. Such tests reveal the uniformity of mixing and the origin of any unusual hot spots on the chamber or the nozzle wall. Transparent two-dimensional motors have also provided valuable information about injector operation.

Combustion. From the standpoint of combustion efficiency, the chamber volume must be made large enough to allow the necessary residence time for the reaction to reach completion, and the length from the injector to the exhaust nozzle must be greater than the ballistic path of the largest droplets before they are consumed. Thus, for each propellant combination and each injector, at a given pressure, the fundamental design specifications are minimum chamber volume and minimum chamber length. A volume larger than the minimum leads to weight, size, and cooling penalties; a length greater than the minimum, for a given volume, means that the convective heat flux in the chamber is unnecessarily high because of the small cross-sectional area.

The residence time in a chamber of volume $\mathcal{U}_c$ is:

$$\tau_c = \frac{\mathcal{U}_c}{\dot{m}/\rho_{av}} \cong \left(\frac{\rho_{av}}{\rho_{min}}\right) \frac{L^*}{c^*} \frac{1}{\gamma} \left(\frac{\gamma+1}{2}\right)^{\frac{\gamma+1}{\gamma-1}} \cong 5 \frac{L^*}{c^*} \qquad (5\text{-}5)$$

The following relations were used in this derivation:
Characteristic velocity (see Art. 2):

$$c^* = \frac{p_o A_{th}}{\dot{m}} \qquad (5\text{-}6)$$

Theoretical c^* (ideal motor theory):

$$c^* = \left(\frac{RT_c}{\mathfrak{M}}\right)^{\frac{1}{2}} \gamma^{-\frac{1}{2}} \left[\frac{(\gamma + 1)}{2}\right]^{\frac{(\gamma+1)}{2(\gamma-1)}} \tag{5-7}$$

Perfect gas law

$$\rho_{\min} = \frac{p_c \mathfrak{M}}{RT_c} \tag{5-8}$$

Definition of "characteristic chamber length"

$$L^* = \frac{\mathcal{V}_c}{A_t} \tag{5-9}$$

The numerical factor at the right side of Eq. 5-5 is based on the values $\gamma \cong 1.20$ and $(\rho_{av}/\rho_{\min}) \cong 2.0$. For a typical rocket motor using oxygen-gasoline, with $L^* \cong 60$ inches and $c^* \cong 5000$ ft/sec, at 300 lb/in.^{2} chamber pressure, the residence time calculated according to Eq. 5-5 is about 5 milliseconds. This time is to be regarded properly as the gas phase part of the total residence time, since the propellant spends a few milliseconds in the liquid phase as droplets, and this would not show up in Eq. 5-5 unless the time-average density ρ_{av} were made very much larger than $2\rho_{\min}$ to reflect the liquid-phase history.

It is the practice to determine $L^*_{\min}$ by measuring c^* for a series of chambers of progressively smaller volume, all at one pressure, and then plotting c^* as a function of L^*. For liquid oxygen-liquid hydrogen at 300 lb/in.^{2}, the minimum is about 10 inches; for nitric acid-gasoline, 50 inches; for nitric acid-aniline, 40 inches; for oxygen-alcohol, 50 inches; and for nitromethane, about 300 inches. Inasmuch as these minima were established by trying various kinds of injectors at different laboratories with different chamber configurations, it appears that they correspond to fundamental chemical reaction times. A calculation of the reaction time for NO decomposition indicates that this may be the step that determines the minimum L^* for nitric acid motors.

This interpretation of L^* suggests that, in scaling up a motor from, say, a 1000-lb thrust laboratory unit to a larger practical one, the proper rule is to increase the volume V_c in proportion to $\dot{m}$, at the same pressure. It has been found that this rule is a good guide. Furthermore, if the over-all reaction kinetics can be assumed to be of second order, the time to completion would vary inversely as the pressure. (See II,D.) According to Eq. 5-5, this means $L^*_{\min}$ can be reduced in the extrapolation of test data to higher pressures. This trend is also confirmed experimentally.

Although the L^* concept determines the volume of the chamber, its proportions have to be decided according to other considerations. It is obvious that the length cannot be shorter than the ballistic path of the larger droplets, that is, the penetration distance. At the present time, experiments to determine the minimum length for any particular injector are a more reliable guide than theory. The longest length for a given L^* is achieved by making the chamber diameter equal to the nozzle throat diameter, but this results in a prohibitive heat flux by convection. The usual practice is to make the chamber diameter about 1.5 to 3 times the nozzle throat diameter. If high frequency instability occurs, modification of the length, either larger or smaller, can stabilize the motor. (See below.)

Ignition. A nonself-igniting propellant combination such as alcohol plus liquid oxygen, or a monopropellant such as ethylene oxide, requires an ignition energy source to get started. There are a number of practical considerations that govern the selection of an igniter: (a) whether one-shot or repeated starts are desired; (b) availability of the energy or substance, for example, electrical power; (c) the necessary degree of reliability (with nitric-acid-gasoline, for example, a misfire can lead to violent explosion); (d) the complexity of the system; (e) the maximum allowable waiting time from the "fire" signal to the moment of ignition, and the desired rate of build-up. It is usually impossible to meet all these specifications to the highest degree simultaneously.

Some practical ignition devices are:

1. Pyrotechnic squibs, which may be mounted in the injector head, the chamber wall, or on an expendable rod pointing forward through the exhaust nozzle.
2. High energy spark plugs, the fastest to start but the most prone to fouling by deposits or short-circuiting by acid spray.
3. Self-igniting chemical starting fuel, for example, xylidene injected just before admission of gasoline, in a nitric acid-gasoline motor.
4. Catalytic surfaces in the chamber, such as ceramic elements saturated with potassium permanganate, for starting hydrogen peroxide decomposition.
5. Auxiliary oxygen, for starting the combustion of JP-4 before admission of the main oxidizer, nitric acid, or for start-up of ethylene oxide.
6. Electrically heated glow plugs, to provide a hot surface that can ignite the initial spray of the mixture (convenient, but slow to start).

There are several special problems connected with ignition that deserve mention. One is the starting of very large motors, 10,000-lb thrust and up. In order to provide a spray that can be ignited, the initial flow rate through a fixed area injector has to be at least about $\frac{1}{4}$ of the rated flow, taking an approximate lower limit. For a larger motor, even this flow may drown the igniter of usual size. A practical solution is to have a starting injector recessed in a small ignition chamber which is connected to the main chamber through a critical flow orifice; the ignition can take place in the small chamber just as in a small motor, and the high intensity flame shooting into the large chamber then ignites the main flow. A pressure-actuated relay makes sure that the main flow is not started until the starting chamber is up to rated pressure.

The start-up of a large motor must be programmed carefully, the usual scheme being to provide for two stages of flow, a warm-up stage of a few seconds, and then rapid rise to rated flow. Without adequate warm-up, there is danger of flame-out by flooding the fire.

A "hard start," or initial pressure overshoot in the chamber, is almost inevitably the result of a delay in ignition, particularly of propellants such as nitric acid-gasoline that can burn explosively or detonate when premixed. Nitric acid reacts immediately upon contact with hydrocarbons to form nitration products that are explosive. If the igniter fails to initiate combustion immediately after the propellants enter the chamber, the sensitive mixture accumulates and explodes almost instantaneously when ignition does take hold. The resulting overshoot in chamber pressure may be sufficient to damage the rocket motor or upset the guidance system or instrumentation in the vehicle.

For the case of an accumulated mixture that is nondetonable, a rough estimate of the maximum allowable delay can be deduced as follows. If the exploded propellant accumulation produces a peak temperature about twice the space average flame temperature for normal constant pressure burning, and if no gas escapes from the nozzle during the explosion, then the mass of gas that would produce a transient explosion pressure equal to the rated pressure is given by the normal flow rate $\dot{m}$ multiplied by one half the normal gas-phase residence time τ_o. Then, using Eq. 5-5,

$$\Delta t_{\text{delay}} \cong 2.5 \frac{L^*}{c^*} \left(\frac{\dot{m} \ (\text{rated})}{\dot{m} \ (\text{start})} \right) \left(\frac{p_{c(\text{trans})}}{p_{c(\text{rated})}} \right) \tag{5-10}$$

If we take $L^* = 60$ in., $c^* = 5000$ ft/sec, $\dot{m}_{(\text{start})} = \frac{1}{4}\dot{m}_{(\text{rated})}$, and $p_{c(\text{trans})} = p_{c(\text{rated})}$, then $\Delta t_{\text{max}} \cong 10$ milliseconds.

It is clear from this result that prompt ignition is essential. Alternatively, as a safety measure, a means must be provided to shut off the propellant and purge the chamber promptly of the accumulated propellant in the event of excessive ignition delay.

The problem of ignition is greatly simplified, particularly for repeated starting, in the case of self-igniting propellants. The most common examples of self-igniting combinations are hydrocarbon amines with concentrated nitric acid. Fuels of this class are aniline (phenylamine), xylidine, hydrazine, unsymmetrical dimethyl hydrazine (UDMH), etc. Furfuryl alcohol, which is not an amine, reacts in a similar fashion, however, with nitric acid. Spontaneous ignition occurs also between fluorine and all of the common fuels. Oxygen ignites upon contact with diborane. Hydrazine and hydrogen peroxide are also self-igniting. It should be observed, of course, that a combination that is spontaneous at room temperature may not ignite at very low temperature, and that a combination that is not spontaneous at room temperature may be self-igniting at elevated temperature. Consequently, the distinction between self-igniting and nonself-igniting propellants is largely one of the degree of reactivity in the liquid phase at room temperature.

From the standpoint of rocket ignition, the appropriate measure of reactivity is the *ignition lag*, that is, the measured time interval between the moment of contact of the two liquids and the subsequent appearance of flame. Such ignition lags lie generally in the range from 5 to 100 milliseconds, but experience has shown that the method of testing can greatly influence the absolute value. Various methods of measuring ignition lag have been used: (1) the open cup test, (2) the closed chamber with forced injection, (3) the falling drop test, and (4) small rocket motors. The values for the identical propellants at the same temperature differ by as much as a factor of ten, depending on the type of test. Nevertheless, the different methods generally agree as to the *order* of reactivity of various propellants and as to the effects of temperature, diluents, and catalysts. Typical ignition lag measurements are given in Table G,5.

Table G,5. Typical ignition delay times for several propellant combinations.

Propellants	Ignition delay, msec
Triethylamine + WFNA	30
Allylamine + WFNA	50
Cyclohexene + WFNA	90
80% Furfuryl alcohol + 20% aniline + WFNA	20
Furfuryl alcohol + WFNA	40

One of the interesting aspects of ignition lag testing is the evidence that the lag depends as strongly on the intimacy of mixing and on the heat insulation as on the chemical kinetics. Thus, forced penetration of nitric acid and aniline by injection results in a much smaller lag. Tests involving large quantities of propellant as compared with small quantities show shorter lags, undoubtedly the result of better heat retention during the first stage of reaction. The addition of "wetting agents" to reduce the interfacial tension between the two liquids and thus promote surface contact has been found very effective: 2 per cent of oleate of polyoxyethylene glycol added to furfuryl alcohol reduced the lag from 25 to 12 millisec in the reaction with nitric acid; certain other substances were similarly effective.

The ignition lag in the rocket motor is an important characteristic of the propellant combination in connection with the "hard start" pressure evaluated in Eq. 5-10. A more complete discussion of the ignition problem from the standpoint of combustion mechanisms can be found in II,L.

Combustion instability. One of the difficult problems of combustor design is that of preventing unstable combustion during the so-called steady thrust period of firing. The phenomenon is observed in all types of liquid propellant rocket motors, but the consequences are more serious the larger the size, and so it has only been in recent years that the problem has become an urgent one. Also, the phenomenon became quite prominent only when part of the liquid rocket engine industry in the U.S. switched strongly over to nitric acid-kerosene-type propellants from the earlier nitric acid-aniline types; it has not been as troublesome in oxygen systems. The reasons for these differences are discussed below.

Instability manifests itself as a strong oscillation in combustion pressure, with amplitudes as large as 50 per cent of the mean pressure, which occurs in the absence of any external stimulation while the control valve positions and feed pressures are absolutely steady. Observed frequencies range from as low as 10 cps to several thousand per second. In the low frequency range, the propellant flow rate oscillates along with the combustion pressure, due to dynamic coupling, but the high frequencies appear only in the combustion chamber because the inertia of the liquid in the lines is too great to follow the rapid pressure changes. The practical consequences of combustion instability are, in the low frequency or "chugging" range, severe vibration of the vehicle and its delicate instrumentation, a reduction in specific impulse, and quite often, rupture of the rocket motor. In the high frequency range, usually called "screaming," the most serious consequence is a large increase in heat flux and usually a burnout of the motor.

Although superficial observations of instability have resulted in a two-fold classification, either "chugging" or "screaming," more refined observations have disclosed at least three basic types of instability, according to the driving mechanism: low frequency, high frequency, and intermediate frequency. The *low frequency type*, below 100 cycles in practical motors, is characterized by a uniform pressure in the chamber which oscillates with time, coupled to a similar oscillation in the propellant flow rate. The basic situation is that of a servomechanism with excessive time delay in the feedback loop, which is therefore unstable. In the case of the rocket combustor, the feedback is either the effect of a p_c fluctuation on the instantaneous $\dot{m}$ (that is, hydraulic or structural coupling with the feed system), or the effect of a similar fluctuation on the combustion reaction rate (chemical kinetics coupling). Both feedback effects may exist simultaneously.

The simplest theory was evolved in 1941 after it was observed that nitric acid motors, which could not be made to burn smoothly on gasoline, did burn smoothly when the tank was filled with aniline. The fact that the former combination was nonself-igniting and not easily ignited, whereas the latter was self-igniting, suggested the idea that "chugging" of the former was due to a slow reaction rate. This concept has been introduced into the theory of instability as the "combustion time lag." In its initial form, the time lag was regarded as a property of the propellant and the conditions of injection, but invariant during the fluctuation of pressure. Therefore, it was not time-dependent. Subsequently it was shown theoretically that important types of instability can occur if the time lag is sensitive to the local instantaneous conditions of pressure and temperature. Since the local temperature during an oscillation is coupled to the pressure in a unique fashion, it is sufficient mathematically to describe this as simply a pressure-sensitive time lag.

The pressure-insensitive time lag is defined by the observation that the rate of evolution of gas at any instant is equal to the rate of injection of propellant at a time τ millisec earlier:

$$\dot{m}_b(t) = \dot{m}_i(t - \tau) \tag{5-11}$$

The interval τ is the time required for the physical processes of vaporization and mixing and the chemical processes of reaction. When this time lag is inserted in the theory of small oscillations of a liquid monopropellant engine with a compressed gas feed system, a simple approximate condition for stability emerges:

$$\frac{l_p \dot{m}}{p_e A_p} + \frac{4.8 L^* \Delta p_p}{c^* p_e} > \tau \tag{5-12}$$

According to this equation, reduction of the time lag τ, either by better mixing and atomization or by substitution of a more reactive propellant for a sluggish one, will tend to stabilize an otherwise chugging engine. (This corresponds to the experience noted above of switching from gasoline to aniline.) Increasing the length of the feed line l_p or the velocity of flow $\dot{m}/A_p$ will also stabilize the system, by increasing the momentum in the feed lines. An increase in L^* produces stability by the capacitance effect of the chamber, and an increase in the injection pressure drop tends to stiffen the flow rate against perturbations. There are experimental results tending to corroborate these predictions. As to the magnitudes of the various terms, the left side is generally of the order of 5 to 10 milliseconds, and time lags have been measured in the range of 1 to 5 milliseconds, indicating the possibility of instability.

The introduction of pressure sensitivity in the time lag (or in part of it) makes instability possible even with a perfectly "stiff" feed system, that is, a system having an absolutely constant flow rate. In this case, the oscillating pressure arises in fact from an *oscillating burning rate*, the coupling being affected by the sensitivity of the time lag. This type of *"intrinsic" low frequency instability* can occur only if the pressure-sensitive time lag is long enough and if the sensitivity is sufficiently strong. The sensitivity to the pressure is described in terms of an interaction index n, which can be defined by saying that the time lag required for the pressure-sensitive process to be completed, if carried out at constant pressure, varies as p^{-n}. The critical value of the time lag for any index n is shown to be:

$$\tau = \left(\frac{L^*c^*}{RT_o}\right)\frac{\left[\pi - \cos^{-1}\dfrac{1 - n}{n}\right]}{[2n - 1]^{-\frac{1}{2}}}$$ (5-13)

This curve is plotted in Fig. G,5a. It shows, first, that no such instability is possible if n is less than $\frac{1}{2}$. Secondly, if the interaction is strong (large n), the motor will easily become unstable. The pressure-sensitive

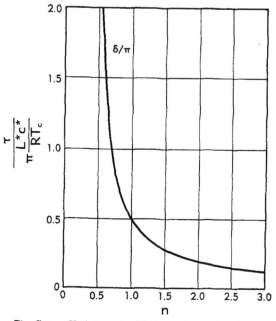

Fig. G,5a.   Variation of critical time lag with index n.

time lag and the interaction index have been measured experimentally, by imposing forced flow oscillations on the pressurized feed system of an oxygen-alcohol rocket engine in stable operation and by observing the phase and amplitude of the chamber pressure as a function of the forcing frequency. Typical values are (at 300 lb/in.² pressure, 1.34 oxygen-alcohol ratio):

$$\text{total } \tau = 0.25 \text{ millisec}$$
$$\text{sensitive } \tau = 0.08 \text{ millisec}$$
$$\text{index } n = 1.0$$

The interaction of the combustion process with the pressure, expressed through the index n, serves to explain also the occurrence of *high frequency oscillations*. The explanation rests on the observed fact that the frequencies are very close to the natural frequencies of the combustion chamber treated as a closed, long cylindrical column of stationary gas. Thus the dominant frequency is usually that of the fundamental axial mode of the "organ pipe." (A two-foot-long chamber filled with typical combustion gas at 3000°K would show a fundamental of about 1000 cycles.) Higher frequencies can be identified with longitudinal overtones, radial modes, and angular modes, all of which can occur

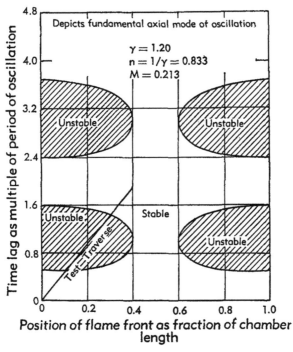

Fig. G,5b. Stability diagram for longitudinal oscillations.

simultaneously. These wave motions would ordinarily decay as a result of viscous dissipation and loss of wave energy through the exhaust nozzle. However, the local speed-up of the exothermic reaction when a compression wave passes through an element of the combustion gas (assuming that $n > 0$) can serve as the mechanism for injecting energy into the wave to compensate for losses. It can be shown that, in order to produce instability of a particular mode, this energy release must take place in regions of the chamber near the pressure antinodes of the stationary wave and it must occur with the right phase lag, about one half the period. These conditions can be expressed, in the case of longitudinal oscillations, by a stability diagram such as Fig. G,5b, in which the unstable values of *space lag* and sensitive *time lag* are defined, for a specified interaction index n. The diagram shows, for example, that the fundamental mode is unstable only if the sensitive part of the combustion process is concentrated either near the injector or near the end of the chamber, with the simultaneous condition that the time lag must have a particular value or a certain larger value (so as to straddle an odd multiple of half-periods). Similar regions of instability are shown for the first overtone.

It is clear that instability can be avoided by choosing an injector type and a chamber length so as to place the sensitive combustion reaction in one of the stable areas of the diagram. Inasmuch as the fundamental is the most powerful mode, the designer must start by ignoring the unstable regions connected with the higher modes. Recent tests of an oxygen-alcohol motor, conducted by varying the length of the combustion chamber while keeping all other con .itions the same, have demonstrated the actual existence of such zones of instability for the fundamental longitudinal mode. This series of tests were carried out along a traverse such as the inclined straight line in Fig. G,5b.

In modern large rocket motors, the most important unstable mode seems to be the transverse "sloshing" mode instead of the longitudinal one. The stability theory for this case has not yet been published, but the principles for avoiding instability are undoubtedly the same.

An important instability with frequencies intermediate between the chugging frequency and the organ-pipe frequencies has been traced to an interaction between pressure waves in the chamber and the mixing or injection process near the injector head. A pressure pulse that starts at the exhaust nozzle moves upstream with the velocity of sound, strikes the injector where it causes unequal fluctuations in the two propellant injection rates or it disturbs the mixing process; the off-mixture layer of propellant is swept downstream toward the nozzle, burning on the way; a new pressure pulse is reflected from the nozzle when this layer (called an "entropy wave") enters the nozzle; in this way, the cycle is repeated. Although an exact theory has yet to be developed for the stability condition, it is evident that the period of oscillation is the sum of the wave travel time and the mean chamber residence time, and that the frequency is therefore intermediate between the organ-pipe fundamental and the chugging frequency.

Other hypothetical instability mechanisms have been suggested but are as yet unverified experimentally: (1) wave propagation in the liquid feed system, supported by oscillations of combustion pressure, (2) oscillatory injection of the propellant, stimulated by elastic deformations of the cooling jacket and the injector manifolds due to the variations in chamber pressure, (3) hydraulic instabilities in injection orifices, (4) spray instabilities due to hydrodynamic forces, (5) self-excited oscillations of the chemical kinetics of consecutive reactions with feedback; and (6) coupling of the rocket engine with a natural vibration of the supporting structure.

Some of these mechanisms are plausible, but it remains for the future to clarify them.

Although the preferred way for a designer to cure a case of combustion instability is to modify the combustor or the operating conditions according to the above theories, some attention has been given to the use of feedback servos to provide automatic stability. The idea is to use the combustion pressure signal to actuate propellant flow controls to modify the injection rate to overcome at least low frequency instability. No such device has yet been put into practice.

Scaling of rocket motors. Scaling is the problem of extrapolating the performance characteristics of one rocket motor, acquired by firing tests, to another rocket motor as yet untested. A narrower version of the problem is to design a large rocket motor that will have the same characteristics as the smaller one that has been successfully developed. Among the characteristics to be preserved or predicted are the combustion efficiency (c^*), the nozzle effectiveness (C_F), the heat flux distribution (q), the margin of combustion stability at both low and high frequencies, the starting chamber pressure pulse, and the working stresses in the important chamber structure.

The present approach of the designer is the obvious one of dealing with each problem separately on the basis of either experimental correlations or theory. The simplest case is that of relating a large motor to a small one, keeping the propellants, their mixture ratio, and the chamber pressure the same. Thus the exhaust nozzle is scaled according to the theory and data in Art. 2 and 3; the heat flux, both convective and radiative, according to Art. 6; the chamber volume according to the L^* concept in Art. 5; the stability margin according to the theories just presented; etc. Incompatible requirements frequently arise: for example, the choice of chamber length is dictated on the one hand by the volume required for combustion together with the cross-sectional area required by convective heat flux considerations, and on the other hand by the condition for stability in the fundamental longitudinal mode. It is not possible at this time to write a universal prescription for such difficulties except to say that the designer must decide to relax one specification or the other, or strike a compromise. The more difficult problem is that of predicting the proper design at a different pressure level as well as a larger size. For this purpose, experimental data are needed on such empirical parameters as c^*, nozzle C_d, time lag τ, interaction index n, radiative emissivity ϵ, etc., all as functions of p_c.

An interesting theoretical approach has been proposed recently, to derive dimensionless similarity parameters from the basic physical equations governing rocket motor performance, just as the Reynolds number emerges from the Navier-Stokes equations for viscous flow. In the case of the rocket motor, regarded as a chemical reactor with heterogeneous chemical kinetics, convective, conductive and radiative heat transfer, steady and unsteady gas dynamics, wall stresses, etc., the problem is a complicated one. This approach is not yet fruitful but is receiving active attention.

Mechanical design and construction. The principles of rocket combustor design, having emerged as a result of experience over the years, can be appreciated only by careful study of a number of actual rocket motors, the early types as well as the modern ones. Many are described in the engineering literature.

Several of the design principles deserve explicit mention:

1. The chamber and nozzle wall and the injector face plate must be made of metals selected for high strength at elevated temperature coupled with good thermal conductivity, resistance to high temperature oxidation, chemical inertness on the coolant side, and suitability for the fabrication method to be employed. In the case of certain monopropellants, the metal must not catalyze the decomposition. Although aluminum and copper alloys have been used successfully for combustion chambers and nozzles, stainless steels and carbon steels are in widest use today.

2. Large scale motors today are fabricated by sheet-forming and welding techniques. Experimental small motors are usually made by machining with the lathe and milling machine, and then assembled with bolts.

3. The chamber wall must be carefully supported to stiffen it against the compressive load of the pressure in the cooling jacket, both during normal operation and after shut-down if the propellant valve is between the cooling jacket and the injector. Similarly, the injector face plate must be supported by stiffening ribs to prevent collapse under pressure.

4. Thermal expansion of the chamber wall must be accommodated during firing, usually by providing an expansion joint on the outer casing. Allowance must also be made for thermal expansion of the injector plate, usually by elastic deformation.

5. Dimensional accuracy of the cooling passages must be assured, during firing as well as in assembly, in order to guarantee cooling performance. Definite spacers must be provided in large, coaxial shell-type combustors. Longitudinal cooling passages are usually used for large motors, but spiral passages are needed for small motors to preserve a reasonable duct width that is not too narrow to guarantee. The ducts are not of uniform width, being usually narrower at the throat than at the chamber. The method of fabrication should assure smooth surfaces in the cooling ducts to minimize the pressure drop in the feed system.

6. The manifold passages in the injector head should be designed to serve also as regenerative cooling ducts with prescribed coolant velocities.

7. The coolant inlet should be at the exhaust nozzle end to take advantage of the lowest liquid temperature at the most severely heated section of the motor.
8. Provision must be made for close alignment of the nozzle to produce a thrust vector accurately perpendicular to the mounting ring.

Finally, it may be remarked that the problems of design are all made more difficult by the need to keep the weight at a minimum. Typical weight figures for modern rocket motors center around 0.01 lb per lb thrust, the ratio being somewhat less for large missile motors and greater for small aircraft motors.

G,6. Cooling of Rocket Motors.

Heat transfer processes in rocket motors. The problem of designing a cooling system for a liquid propellant rocket motor can be divided into two parts. The first is an examination of the expected magnitude of the heat flux into the walls, based on conditions inside the combustor; the second is an analysis of cooling schemes capable of keeping the wall temperature below the structural failure point.

Two processes account for the heat flow from the combustion gas to the wall, convection and radiation, and the heat flux is the sum of these two contributions. Each depends in a complicated way on the flow velocity distribution, the composition distribution, and the temperature distribution inside the combustor and the nozzle, and these distributions are never known or predictable except in a very rough way. Consequently, precise heat flux predictions are not feasible. The most the designer can hope to calculate is a probable upper limit of the heat flux, or if this is known at one pressure for a given rocket motor, he can make an estimate of the variation of heat flux with chamber pressure or mixture ratio or size.

The cooling of rocket motors can be accomplished in a variety of ways. In this discussion, the meaning of "cooling system" is generalized to include any scheme designed to limit the wall temperature, even if no fluid coolant is employed. Thus the most direct way to limit the internal surface temperature is to provide a sufficiently thick chamber or nozzle wall with the necessary heat capacity to soak up the heat transferred during the prescribed firing duration. This method has the advantages of simplicity and cheapness of manufacture, but for durations greater than 10 to 20 seconds it results in a weight penalty. For longer durations, a better solution is to use one or both of the propellants as coolants ("regenerative" cooling) with lightweight, thin metal, combustion chambers and nozzles. Either the heat capacity motor or the regeneratively cooled motor can be aided by internal protective techniques, such as: (1) refractory insulating nonmetallic liners or inserts to reduce the heat flux, (2) internal coatings applied either as a paint before firing or by continuous deposition from the combustion gas to the wall, (3) liquid or gaseous films on the inner surface supplied by continuous injection through either a porous wall or small orifices in the wall, and (4) low temperature gas or even nonburning propellant in the region near the wall, supplied by appropriately modified injector orifices.

For the practical designer, the following subarticles on heat transfer from the hot combustion gas to the wall by convection and radiation may be less useful than that dealing with cooling methods, since the usual practice is to test a water-cooled experimental motor of the proposed configuration first, and to accept the measured heat flux distribution as the basis of the cooling system design. The study of the high temperature side is useful only for making predictions of heat flux in motors not yet tested and still being designed.

It is not possible in the space of this article to develop all of these processes of heat transfer and cooling in full detail. Instead, only the main principles are presented, and references are given to more complete treatments. The observed heat transfer distributions are discussed first, since they serve to illuminate the more theoretical material to follow.

Observations of heat transfer in experimental rocket motors. The heat transfer characteristics of a rocket motor at a particular operating condition are most conveniently presented in terms of longitudinal and circumferential distributions of the heat flux per unit area. This presentation is useful only because the flow of heat is practically independent of the manner of cooling the chamber. This is so because the inner wall temperature of a cooled motor must be kept so much lower than the flame temperature that differences in wall temperature due to different cooling methods cause only small differences in heat flux. If the heat flux distribution were sensitive to the particular conditions of the cooling system, it would be necessary to present the results in terms of conductance coefficients, as in conventional heat exchanges.

Measurements of heat flux distribution are made in several ways, but the most convenient method makes use of a sectional, water-cooled rocket motor. The internal geometries of the combustion space, the injector, and the nozzle are made identical with the practical motor. In the test motor, each section is cooled by water, and the *average* heat flux in the section is calculated from the increase in water temperature from inlet to outlet. If measurement of *local* heat flux is desired, it is possible to position two thermocouples at different radial distances from the inner surface, and on the assumption of purely radial heat flow, the heat flux is calculable from the temperature difference and the thermal conductivity of the wall material. One of the simplest methods is to use a "heat capacity" motor, for example, a rocket motor made of sections of solid copper, one insulated from the other; the average heat flux can be calculated from the transient rate of temperature rise of each section. Unfortunately, the methods based on the use of sectional motors result in circumferential averages of the heat flux and fail to indicate variations in that direction. Such circumferential variations may be important if the injector is not exactly axially symmetric, and most are not. The most obvious experimental clues to the existence of heat flux peaks in the circumferential direction are longitudinal discolorations (streaks) of the inner surface of the chamber after a firing. Such hot zones are frequently visible in the exhaust jet also.

In the evaluation of a measured heat flux distribution, it is important to realize that two motors operating on the same propellants, at the same mixture ratio and at the same pressure, can exhibit different heat fluxes by a factor of two, because of various unpredictable effects. Thus, in a nitric acid motor at 300 lb/in.^{2} pressure, it was found that an annular spray-type injector produced 36 per cent higher average heat flux (averaged over the entire inner surface) than an impinging jet injector, although the specific impulse was increased only 5 per cent. A similar sensitivity to injector design is shown by the fact that a slight change of about 5° outward in the direction of the mixed streams produced by pairs of impinging jets can result in the burnout of an otherwise satisfactory motor. As a practical technique, it is possible to design a large injector plate so as to produce a relatively cool layer of combustion gas at the wall, while the bulk of the injected propellant burns at high temperature, and thereby reduces the heat transfer. Rough or unsteady combustion generally results in higher heat fluxes in the chamber, and under the extreme condition of "screaming" combustion, the heat flux in the same motor has been found to increase by more than 100 per cent. Another point well known to rocket designers is that any break in the smooth contour of the inner surface can raise the heat flux locally; such breaks are made by a pressure measurement orifice, a sharp nozzle throat, or a turbulence promotor or flame holder in the combustion chamber. Finally, it must be observed that solid deposits formed on the walls, such as carbon, iron oxides (from iron dissolved in nitric acid in steel tanks), boron oxides, etc. tend to reduce the heat flux through their insulating effect. This phenomenon has been used intentionally to cool a motor, by adding a few per cent of a silicone oil to the fuel, to produce a protective deposit of silica.

If one considers the longitudinal distribution of the circumferentially averaged heat flux in a typical nitric acid rocket motor, some noteworthy features are: (1) the order of magnitude of heat flux, 1 BTU/in.^{2}-sec, about 10 to 20 times that of a high-output, aircraft-type combustion heater, (2) the high flux in the vicinity of the throat, due mainly to strong convective heat transfer, (3) the sharp fall-off after the throat, due to the expansion cooling and the build-up of the boundary layer, and (4) the slow rise along the combustion chamber wall corresponding to the axial progress of combustion.

Considering the effect of mixture ratio on the heat flux in a rocket motor, one finds that the peak values of heat flux occur near the peak flame temperature. For hydrocarbon fuels, this occurs on the fuel-lean side of the ratio for peak specific impulse, as explained in Art. 4.

The effect of pressure on the heat flux in a nozzle is in general an increase in the heat flux with increasing pressure due to a progressive increase in boundary layer conductance with increasing gas density.

Finally, it should be observed that a typical value of the integrated heat flux over the entire surface, for a well-designed, 1000-lb thrust motor operating at 300 lb/in.² on either nitric acid-hydrocarbon or oxygen-hydrocarbon, is about 2 per cent of the heat of combustion. This percentage is considerably smaller for larger thrust motors, because the combustor volume increases approximately in proportion to the mass flow, while the surface of the combustor increases only as the two-thirds power, for similar shapes. Thus the heat transfer in the V-2 motor represents about 0.7 per cent of the heat of combustion.

The practicability of running a rocket motor continuously with only regenerative cooling, using either one or both propellants, is directly the result of this low percentage of heat transfer. None of the common propellants, such as JP-4 or alcohol, can absorb more than a few per cent of the heat of combustion without vaporizing or decomposing and thus becoming unsuitable as coolants. This observation can be presented another way, that for each propellant combination and combustion pressure, there exists a certain minimum thrust rocket motor below which simple regenerative cooling is inadequate. Fortunately, for the common propellant combinations based on nitric acid or oxygen, this is well below 1000-lb thrust. For propellants that burn slowly and require large residence times, such as nitromethane, a simple regenerative motor may not really be practical except for a thrust in excess of, say, 10,000 lbs.

Convective heat transfer from the combustion gases to the wall. Convection generally accounts for the main part of the heat flow from the hot combustion gases to the walls of the rocket motor, usually two-thirds or more, depending on the particular location of the surface element in question and the temperature and composition of the gas mixture. Unfortunately, there is no firm basis for theoretical prediction of the heat flux, although several useful empirical correlations of restricted applicability have been developed. The main difficulty is that the temperature and gas flow patterns inside the rocket motor seem to vary strongly in an unpredictable manner from one injector to another, even in some instances where the configurations are similar and the specific impulses are nearly the same. Thus, the practical engineer refers to "hot" injectors and "cool" injectors, the latter class being preferred. Another obstacle to theoretical treatment is that there is not yet sufficient information on the boundary layers involved, e.g. at the throat of a cooled de Laval nozzle. Finally, the transport properties of the gas mixture, viscosity, and thermal conductivity are not well known at the elevated temperatures involved.

Despite these uncertainties, empirical correlations have emerged which can be applied to determine within about ± 20 per cent the convective heat transfer, although the "hot spots" produced characteristically by "hot" injectors are not predicted by these methods. The discussion can be divided into three parts dealing, respectively, with the subsonic cylindrical combustion chamber, the supersonic portion of the de Laval nozzle, and the throat of the nozzle.

In the chamber, strongly nonsteady flow conditions prevail, and so the boundary layer is undoubtedly turbulent. It has been found that the heat transfer at the hot end of the chamber can be correlated by an equation of the Colburn type in which the pertinent length is the chamber diameter d_c:

$$Nu_d = aRe_d^{0.8}Pr^{0.33} \tag{6-1}$$

where Nu_d is the Nusselt number, Re_d is the Reynolds number, and Pr is the Prandtl number defined as follows:

$$Nu_d = \frac{hd_c}{k}$$

$$Re_d = \frac{\rho V d_c}{\mu} \tag{6-2}$$

$$Pr = \frac{\mu c_\rho}{k}$$

The numerical coefficient a has the value 0.023 in fully developed turbulent pipe flow, but there is no a priori reason to apply to the short-tube rocket combustor the same numerical coefficient. Nevertheless, test results in rocket chambers indicate that Eq. 6-1 gives a reasonable estimate of the convective heat transfer coefficient h, with about the same numerical coefficient. This applies particularly to the downstream end of the cylindrical combustion chamber.

The transport properties μ and k are sensitive to gas temperature, and so it is necessary to specify the reference temperature at which they are to be evaluated for use in Eq. 6-2. The best correlation, at least in the case of pipe flow, is obtained by using a reference "film" temperature

$$T_f = T_w + 0.5(T_g - T_w) \tag{6-3}$$

where T_w is the wall temperature and T_g is the local gas temperature. Use of this "average" temperature seems justified also for the rocket motor. Fortunately, because the Prandtl number is nearly independent of temperature and because h depends only on the 0.2 power of μ, the sensitivity of the result to the reference temperature is not very great.

(For a general discussion of the dependence of transport properties on gas temperature, consult I,D.) Finally, it is general practice to evaluate ρ and c_p at this reference temperature, too.

The heat flux from the hot gas to the wall becomes:

$$q = h(T_s - T_w) \tag{6-4}$$

In the supersonic portion of the deLaval nozzle, as in the combustion chamber, the flow conditions are not like those of the classical flat plate or straight cone which have been the objects of many researches. Therefore it is necessary to rely heavily on the few systematic nozzle tests that have been performed. It appears that, in rocket nozzles of practical size, the boundary layer in the major portion of the diverging section is turbulent, and that the proper Reynolds number for the situation is based on the distance L downstream of the throat. The boundary layer at the throat is very thin, as a result of the steep negative pressure gradient, and so the boundary layer downstream may be considered to have started at the throat as a leading edge. By analogy with the flat plate in a uniform supersonic stream, a correlation of the following form may be expected for the local heat transfer coefficient:

$$Nu_L = aRe_L^{0.8}Pr^{0.33} \tag{6-5}$$

The numerical factor a has the value 0.0296 for the case of a turbulent boundary layer on a flat plate with zero pressure gradient and small temperature difference. For the divergent cone of a Laval nozzle, experiments indicate values of a ranging from 0.025 to 0.028. Like the case of the flat plate in supersonic flow, the heat transfer coefficient h that enters into Nu is defined in terms of the adiabatic wall or recovery temperature T_r, thus

$$q = h(T_r - T_w) \tag{6-6}$$

This relation serves to define also T_r, namely, the local temperature that would be created on the surface of a wall made of nonconducting material. In low speed incompressible flow, T_r would equal the static temperature of the gas T_s but, in high speed compressible flow, T_r is more nearly equal to the stagnation temperature of the gas T_s^0. This effect, sometimes referred to as frictional heating, is expressed in terms of a temperature recovery factor r:

$$r = \frac{T_r - T_s}{T_s^0 - T_s} \tag{6-7}$$

In steady, adiabatic flow it is impossible for a gas with Prandtl number less than unity to develop a temperature anywhere that exceeds stagnation temperature, so that r must be less than unity. For compressible, turbulent boundary layers along flat plates, wedges, and cones, with zero or near-zero pressure gradient, experiments in the super-sonic range up to $M = 4$ indicate that:

$$r \cong Pr^{\frac{1}{3}} = 0.90 \text{ (for air) and } 0.93 \text{ (for CO}_2) \qquad (6\text{-}8)$$

A mixture of diatomic and polyatomic gases, such as that which occurs in rocket combustion, would exhibit a recovery factor of about 0.91.

The gas properties, Pr, c_p, μ, k, and ρ vary with temperature, par-ticularly the transport coefficients. It is therefore necessary to specify a reference temperature at which they are to be evaluated. The best correlation of the experimental results has been achieved with the follow-ing reference temperature:

$$T_t = T_w + 0.23(T_g - T_w) + 0.19(T_r - T_w) \qquad (6\text{-}9)$$

This relation converges approximately to Eq. 6-3 as M goes to zero.

For the laminar compressible boundary layer, the defining equations (Eq. 6-6 and 6-7) are still employed, but the following relations for Nu and r replace Eq. 6-5 and 6-8:

$$Nu_L = 0.33 Re_L^{0.5} Pr^{0.33} \qquad (6\text{-}10)$$

$$r \cong Pr^{\frac{1}{3}} = 0.86 \text{ (for air) and } 0.90 \text{ (for CO}_2) \qquad (6\text{-}11)$$

Eq. 6-10 is the same correlation as that for flat plates in incompressible flow with zero pressure gradient. The effects of compressibility enter in the use of T_r instead of T_g in Eq. 6-6, the use of r to obtain T_t, and the use of Eq. 6-9 for T_f instead of Eq. 6-3. It is possible that the correla-tions for a laminar boundary layer should be applied to the region of the exit cone near the throat, but there is not sufficient experimental informa-tion to support this idea. It is current practice to apply the turbulent correlations throughout the exit cone.

A remarkable fact, however, is that the heat transfer at the nozzle throat is *also* predicted with fair precision by an equation like Eq. 6-5 in which the significant dimension seems to be the diameter d_{th} of the throat rather than some length along the surface. This equation has been found to hold for a throat having a contour radius approximately equal to the throat diameter; it would not apply to "sharp" throats:

$$Nu_d = 0.023 Re_d^{0.8} Pr^{0.33} \qquad (6\text{-}12)$$

This is exactly the same as Eq. 6-1, the formula for incompressible flow in tubes. However, because of compressibility, the gas properties are evaluated at T_t, as in Eq. 6-9, and T_r is introduced in the heat flux equation, as in Eq. 6-6. The recovery factor can be taken equal to unity, with small error.

In summary, there is some experimental support for the utilization of the correlations just presented to predict the convective heat transfer at the downstream end of the cylindrical chamber, at the nozzle throat, and in the exit cone. From the viewpoint of boundary layer theory, there is considerable basis for misgiving when correlations that have been established for the idealized cases of the flat plate and the infinite tube are used for the much less ideal rocket motor. All that can be said in defense of such practice at the present time is that it gives reasonable results. Future investigations may very well disclose the existence of some coincidences that enable a doubtful treatment to produce acceptable results.

In the application of the preceding correlations, it is necessary to evaluate V, ρ, μ, k, c_p, T_s, T_s^0, T_r, T_w, and T_f. On the basis of theoretical combustion calculations (Art. 4), V, T_s, and T_s^0 are known. Some accuracy can be gained by adjusting T_s^0 by means of the factor $(c_{exptl}^*/c_{theo}^*)^2$ to allow for combustion inefficiency and a consequent reduction in the actual flame temperature. Correspondingly, V and T_s can be adjusted. The surface temperature T_w is either prescribed in the structural design analysis or determined by the over-all heat transfer analysis to be described later. With very little error, the recovery factor r can be regarded as independent of temperature and assigned a fixed value, and thus T_r is determined. With T_s^0, T_s, T_w, and T_r known, T_f can be calculated. Two questions arise in the evaluation of the proper physical properties: how to deal with a mixture of gases and how to allow for the effect of high temperature.

Inasmuch as data are generally lacking on the transport properties of mixtures, it is common practice to treat each of the properties as a molar weighted average of the pure gas properties. Thus,

$$k = \frac{\sum n_i k_i}{\sum n_i} \tag{6-13}$$

It is known that serious exceptions to this rule exist, but when specific data are lacking, this formula is probably the best to adopt. A similar average is recommended for μ, the viscosity of the gas mixture. (These problems are discussed in I,D.) The specific heat c_p is a molar average, quite rigorously

$$c_p = \frac{\sum n_i C_{pi}}{\sum n_i m_i} = \frac{\bar{C}_p}{\bar{m}} \tag{6-14}$$

Similarly, the mixture density is a molar average of the component densities at the same temperature and total pressure:

$$\rho = \frac{\sum n_i \rho_i}{\sum n_i} \tag{6-15}$$

The preceding treatment is appropriate for heat transfer without dissociation in the main flow outside the boundary layer, that is, for the case in which the composition everywhere in the boundary layer is the same as the main stream.

The effects of elevated temperature on ρ and c_p are well known and require no discussion here. The viscosity μ increases with the absolute temperature somewhat more rapidly than the $\frac{1}{2}$ power predicted by gas kinetic theory for hard-sphere nonattracting molecules. For a combustion gas mixture, the viscosity is nearly proportional to the first power of the temperature. Less is known about the effect of temperature on thermal conductivity, but the Eucken relation which is based on the theory of heat conduction indicates that the Prandtl number should be nearly independent of T:

Eucken (1913): $\qquad k = \frac{1}{4}(9\gamma - 5)\mu c_v$

$$\left.\begin{array}{l} \\ \\ \end{array}\right\} \quad (6\text{-}16)$$

equivalent to: $\qquad Pr = \dfrac{4\gamma}{(9\gamma - 5)}$

From the variation of μ and c_p with T, it is possible then to deduce the variation of k.

With regard to the effect of dissociation on the convective heat transfer, theoretical analysis indicates that, if the Lewis number Le for the gas mixture is near unity, that is, the molecular diffusivity equals the thermal diffusivity, and if the dissociative composition corresponds to the equilibrium value at each point in the cooled boundary layer, that is, recombination takes place fast enough to keep up with the diffusion of the gas toward the wall, then the heat transfer can be correctly calculated by means of the conventional formulas for nonreactive air, provided the specific heat is modified to include the energy of dissociation. In other words, this statement is that the driving enthalpy difference $c_p(T_s - T_w)$ is to be replaced by $(h_s - h_w)$ so as to include the dissociation enthalpy. Surprisingly, it makes practically no difference if it is assumed that the recombination process is too slow to be effective, that is, if the boundary layer composition is frozen, provided that total recombination occurs at the wall. This is so because the transport of energy by free atom diffusion is just as effective as transport by thermal diffusion, if Le is unity. But, if the composition in the boundary layer is frozen, and if the wall is totally noncatalytic so that no recombination occurs anywhere, the heat transfer rate is reduced. In this case, the correct formula makes use of the average specific heat of the frozen gas mixture, and the driving enthalpy difference is simply $c_p(T_s - T_w)$.

It is general practice to ignore the contribution due to recombination in the boundary layer or at the wall, since the additional heat transfer is well within the range of uncertainty of these calculations, for the common propellants at optimum specific impulse mixture ratios. It should probably be included in such high performance, highly dissociated propellants as oxygen-hydrogen.

Radiative heat transfer in rocket combustion chambers. The high temperature combustion gas mixture in the chamber can radiate strongly to the internal walls of the combustor, and there are some experimental results that indicate that such heat transfer can amount to as much as 25 per cent or more of the total heat flux at the wall. In general, the radiative heat flux increases with increasing temperature, with increasing pressure, and with increasing chamber diameter. Polyatomic gases radiate more strongly than diatomic gases, so that, for example, the products of hydrocarbon combustion (CO_2, H_2O) radiate more energy than the decomposition products of hydrazine monopropellant (N_2, H_2). Incandescent particles such as carbonaceous soot and light metal oxides produced in the combustion process may easily contribute the largest fraction of the radiative heat flux. Chemiluminescent radiation from the active reaction zone, observable mainly in the visible and near ultraviolet, seems to be a negligible part of the heat flux, in comparison with the thermal radiation emitted by the gaseous and solid reaction products.

It is necessary at the outset to define precisely the several types of radiation that may be encountered in a combustion chamber, and for this purpose we rely heavily on the extensive theoretical and experimental research carried out by physicists since the latter part of the nineteenth century. Thermal radiation is emitted by a hot body (solid, liquid, or gaseous) merely by virtue of its temperature. Radiation can be stimulated by nonthermal processes as well, such as electron bombardment, X-ray absorption, etc. In a combustor, an important nonthermal stimulus for radiation is chemical reaction; the radiation in this case is called chemiluminescence. With regard to thermal radiation, it is usually necessary for calculation purposes to assume that equilibrium conditions prevail, that is, that the temperature of the gases, of the enclosure walls, and of the contained radiation are uniform and equal. In the literature, the term "thermal radiation" usually is intended to imply the existence of such equilibrium. In a practical combustor, however, radiative equilibrium cannot exist, because of the relatively cold walls; but because the radiated energy is only a small fraction of the thermal energy of the gas, the thermodynamic equilibrium in the gas is not appreciably upset, and equilibrium theory may be used for calculating the radiation intensity emitted by the gas. A similar justification allows the use of Kirchhoff's law in practical heat transfer calculations to equate the spectral absorptivity of a combustion gas to its emissivity. In this case, however, it is more correct to evaluate the total absorptivity at the temperature characteristic of the incident radiation rather than at the temperature of the gas.

Consideration of thermal radiation starts most conveniently with so-called black body radiation. In a strict sense, black body radiation can be produced only inside a sealed cavity with uniformly heated containing surfaces. For experimental measurement of the characteristics of the black body radiation, a heated cavity is used as a source, except that a very small aperture allows a beam to reach the spectrophotometer. The spectral energy distribution as a function of temperature is described by Planck's equation, and the integrated energy flux is given by the Stefan-Boltzmann equation. The practical significance of the concept of black body radiation is that the emissive power of a black body (the aperture in the heated cavity) cannot be exceeded by that of any material or surface in any spectral interval if the excitation is purely thermal. This means that the spectral emissivity of any heated body is less than unity, and so is the total emissivity.

A "grey body" or a "grey" emitter is one that radiates, when heated, an amount of energy in each spectral interval that is a uniform fraction of that of a black body of the same temperature, that is, a grey emitter has a uniform spectral emissivity, which is therefore equal to its total emissivity. No substance is known that is strictly grey, but many opaque solids display nearly constant emissivity over a wide spectral range.

Of particular interest for this discussion of the rocket combustion chamber is the thermal radiation emitted by a cloud of carbon particles or soot that may be formed by the combustion of a carbonaceous fuel, especially with a fuel-rich mixture. Such flames have been called "luminous flames" to distinguish them from the nonsooty flames that radiate very little in the visible region of the spectrum. The emissivity of a luminous flame depends on the type of particles, the size distribution, and particularly on the optical thickness, that is, the number density of the particles multiplied by the path length in the line of sight. Luminous flames with small numbers of particles are generally nongrey, that is, their spectral emissivities are usually greater in the blue than in the red end of the spectrum, and this must be taken into account in the evaluation of the total emissivity of the flame. However, if the particle density is sufficient to produce a total emissivity of 0.2 or more, the contribution to the emission spectrum due to the luminous soot or ash can be regarded as "grey." This contribution is, of course, superimposed on the non-luminous emission of the flame. The luminous emissivity can easily attain values of 0.7 to 0.9 in fuel-rich rocket combustors, and probably also in chambers fed with such fuels as the borohydrides that would produce a smoky ash. In the range in which grey emission holds, the emissivity increases with increasing optical thickness, asymptotically approaching unity by an exponential formula (Beer's Law).

The nonluminous radiation emitted by the hot combustion gas consists mainly of the characteristic vibration-rotation and pure rotation bands of the various polyatomic and heteropolar diatomic molecules. These gases include, in the approximate descending order of their emissive powers: CO_2, H_2O, CO, NO, OH, and HF. The bands all appear in the infrared region of the spectrum ($\lambda > 1\mu$) and are therefore not visible; that is, the gases are transparent and colorless. At temperatures above 5000°K, the ultraviolet electronic bands of these and other gases would probably make significant contributions to the total emissivity, but there are no experimental data for such temperatures. Homopolar diatomic gases, such as N_2, O_2, H_2, and F_2, emit no radiation in the infrared or visible, and therefore contribute nothing to the total emissivity of the gas mixture at temperatures less than 5000°K.

Unlike the radiation emitted by a "grey" sooty flame, the radiation from a nonluminous flame is concentrated in relatively narrow spectral bands, which in turn consist of groups of closely spaced spectral lines. At very high pressures, above 100 atmospheres, the bands tend to broaden, but at the combustion pressures of liquid rockets, the band envelope is substantially the same as at atmospheric pressure. While the total emissivity of a sooty flame does not vary strongly with either pressure or temperature if the same optical thickness (particle density times path length) is considered, the pressure and temperature effects may be quite substantial in the case of banded spectra. The effect of temperature on the total emissivity at constant optical thickness is a consequence of three distinct physical processes: (1) the envelope of an emission band broadens as the temperature is raised because the molecular system is excited statistically to higher energy levels; (2) the maximum of the Planck curve shifts to shorter wavelengths (Wien's displacement law) as the temperature is raised, while the spectral regions of band emission occupy correspondingly weaker sections of the long wavelength end of the Planck curve; (3) the individual rotational lines, broadened mainly by intermolecular collisions, become narrower as the temperature is raised while the pressure is held constant. The first process tends to raise the emissivity of a gas, while the last two reduce it, as the temperature is raised. The second process is usually the dominant one above 1500°K.

The effect of pressure on the emissivity at pressures below 100 atm results from the collision broadening of the rotational lines: it increases with increasing pressure, but reaches a saturation limit asymptotically when the lines have become smeared out so as to overlap each other. For polyatomic molecules such as CO_2 and H_2O, the saturation occurs near atmospheric pressure; for CO and OH at somewhat higher pressures; but the effect is probably "saturated" at rocket combustion pressures.

Other complications that must be considered in the calculation of hot gas emissivities are: (1) the departure of the total emissivity from exact exponential dependence on optical thickness, a consequence of the line character of band emission, (2) overlapping of bands in the spectra of the various gases in a mixture, that is, nonadditivity of the component emissivities, and (3) similar nonadditivity of the thermal emissivities of the nonluminous gases and the luminous soot or ash.

Finally, chemiluminescent radiation should be considered. As stated earlier, the radiation in this category is produced mainly in the active reaction zone, and arises from excited intermediates such as CH, C_2, HCO, and CO_2. Quantitative measurements of the energy emitted by these species in typical hydrocarbon-air flames indicate that this can be ignored as a source of heat transfer; the total radiated energy in the visible and ultraviolet bands is of the order of 10^{-7} of the heat of combustion, as observed at 1 atm pressure. Because of the nonthermal origin, it is not surprising to find experimentally that such bands can exhibit spectral emissivities greater than unity and rotational or vibrational temperatures higher than the gas temperature. Such experiments indicate that the relative intensities of these bands drop as the pressure is raised. There do not seem to be any definitive measurements on the intensity of chemically excited radiation that might be emitted in the infrared, particularly by the much more abundant stable products such as CO_2, CO, and H_2O; it is the usual practice to ignore such radiative heat transfer on the assumption that it is small.

The preceding paragraphs represent a brief survey of the nature of the radiation that is emitted by the combustion gas in a rocket chamber. For engineering purposes, it is desirable to express the complex radiative character of the gas in terms of an emissivity factor which is then to be inserted in the heat transfer equation below to calculate the heat flux. It is evident from the foregoing discussion that there are still many unknown factors, so that experimental heat transfer measurements must be used as a guide in calculating emissivities. Additional complications that require consideration are: (1) the unknown absorption coefficient of the internal combustor wall for the particular radiation emitted by the gas, (2) the uncertain temperature and composition distribution within the chamber and the difficulty of handling theoretically the radiation from a nonuniform gaseous emitter, (3) the unknown particle density of carbon soot or ash, and (4) the complicated geometrical shape factors for the rocket combustor.

Under the circumstances, it has been the accepted practice to calculate the radiative heat transfer by making a number of broad simplifying assumptions. Surprisingly, the few experiments that have been made on the radiated inside rocket chambers have not disagreed seriously with the calculated results. For rockets operating on C-H-O-N propellants at pressures not over 50 atmospheres, these assumptions are:

1. The only important emitters are CO_2, H_2O, and CO, and these radiate thermally.
2. The gas temperature is uniform and equal to $(c^*_{\text{exptl}}/c^*_{\text{theo}})^2 T_{c(\text{theo})}$.
3. The gas composition is uniform and corresponds to equilibrium at the theoretical temperature.
4. The beam length for a cylindrical combustor is 0.9 times the diameter.

5. The component emissivities of CO_2 and H_2O are the same as those at 1 atm pressure, for the same temperature and optical density.
6. The emissivity of CO for a given optical path length is 0.5 of that of CO_2.
7. The correction for overlapping of the H_2O and CO_2 spectra is the same as that at 1 atm.
8. The absorptivity of the chamber wall is unity, that is, it is a black receiver.
9. The wall temperature is so low that back radiation is unimportant.
10. The intensity of the chemiluminescent radiation is negligible for heat transfer.
11. If soot is not observable in the rocket exhaust, the luminous component of the emissivity may be neglected.

With these assumptions, it is possible to make use of the emissivity data that have been compiled by Hottel and his collaborators. The necessary data for calculating the emissivity of a hot gas mixture of CO_2, H_2O, and CO are available in V,I. The heat flux per unit area into the chamber wall is then given by

$$q = \epsilon_g \sigma T_g^4 \tag{6-17}$$

Cooling methods. At the beginning of this article, a number of approaches to the design problem of cooling a rocket motor are mentioned, which need to be analyzed in greater detail. It is necessary to decide at the outset whether the simple but limited duration heat capacity motor or the more complex, unlimited duration, regeneratively cooled motor is most suitable for the intended application. For liquid propellant boosters, for short-duration aircraft JATOS (especially if jettisonable), and for anti-aircraft guided missiles such as NIKE, the heat capacity type is usually adequate; for auxiliary aircraft power plants and for missile sustainer motors, the regenerative type is best. In either case it is possible to employ one or more of the internal protective schemes mentioned earlier, such as refractory liners, refractory deposits, film cooling, sweat cooling, or peripheral low temperature combustion gases, as a supplement to the basic cooling scheme.

The only scheme to be discussed in detail in this article is straight regenerative cooling. References are given at the end to the published literature where full treatments can be found for the cooling methods not discussed here.

The basic idea of regenerative cooling is that one of the propellants, the fuel, is used as a coolant. Since the most intensely heated region is the exhaust nozzle throat, it is best to have the cooling jacket inlet at the nozzle end rather than at the chamber end. For a coolant such as JP-4 or kerosene, it may be assumed that there is a maximum allowable temperature, about 350°F, above which decomposition or coking may take place in the fuel jacket, with the resulting deposition of a harmful insulating carbonaceous layer on the surface to be cooled. It is necessary to avoid raising the fuel above this temperature anywhere in the cooling duct. For any coolant, there is a maximum allowable temperature: for hydrazine or nitromethane, it would be the explosion temperature; for liquid oxygen or nitric acid, or for any coolant, the boiling temperature might be the maximum. To determine whether the proposed design will allow the danger temperature to be reached, it is necessary to make temperature calculations for at least three points in the cooling duct: (1) the mixed stream at the outlet of the duct, (2) the wall at the nozzle throat, and (3) the wall at the injector end of the chamber. Other suspicious places may be indicated by the results of these three calculations.

The temperature of the fuel stream at the outlet is determined by the rate of fuel flow and the total heat flux integrated over the entire cooled area of the combustion chamber and nozzle. (For this discussion, the cooled injector head is included as part of the cooled chamber.)

$$T_{liq2} = T_{liq1} + \frac{\int_1^2 q \, dA}{\dot{m}_f c_f} \tag{6-18}$$

If T_{liq2} exceeds the boiling point of the fuel *at the pressure in the combustion chamber* (p_c), trouble will occur in the form of unsteady propellant flow due to bubbling in the fuel stream in the injector orifices. If T_{liq2} is high enough to exceed the boiling point *at the pressure at the outlet of the cooling jacket*, momentary "vapor-lock" of the fuel system will occur and result in either burnout or severe chugging of the motor. Inasmuch as the integrated heat flux is practically insensitive to the degree of cooling of the wall, as mentioned earlier, the value of T_{liq2} cannot be reduced by modifying the cooling duct but only by a change in the combustor or by use of one of the internal protective schemes, such as film cooling.

The calculation of the wall temperature on the liquid side at any station, such as the throat th, can be discussed by means of Eq. 6-18. With the subscript $_2$ replaced by subscript $_{th}$, it is possible to calculate the bulk temperature in the cooling duct at station th; the integration of q between the two limits is then performed. Then, with the heat flux at th given by experiment, and with the heat transfer coefficient h_{liq} calculated by a formula similar to Eq. 6-1, the wall temperature on the liquid side $T_{w,liq}$ can be calculated:

$$T_{w,liq} = T_{liq,th} + \frac{q_{liq,th}}{h_{liq,th}} \tag{6-19}$$

The evaluation of q_{th} and h_{liq} is not simple to carry out. The magnitude of q_{th} in Eq. 6-19 is usually less than the value given in the case when

q is the heat flux per unit area *on the gas side surface*, while the value of q_{liq} is based *on the effective area being cooled on the liquid side*. The two values of q are in the inverse ratio of the heated area to the effective cooled area. The heated area is usually easy to determine; the effective cooled area depends on the design of the cooling duct. The only simple case of duct design is that of the annular duct with longitudinal or spiral flow. Designs incorporating spiral tubes or fins are more complicated, but solutions of such problems can be found in the usual heat transfer literature.

The film conductance h_{liq} can be calculated for a formula such as Eq. 6-1 with the reference temperature given by Eq. 6-3. In all practical cases, this formula is justified because the flow is in the fully developed turbulent range. However, the coefficient a has been evaluated experimentally only for simple shapes such as flat plates, straight round tubes, and two-dimensional ducts. Coiled ducts, especially of noncircular cross section, spiral passages, finned passages, etc., all of them useful in rocket-cooling jacket design, have not been fully investigated, and probably never will, particularly under a special condition of heat flow which is not uniformly distributed around the surface of the passage. Although it is possible, in principle, to analyze such problems or to perform systematic model tests, the designer usually resorts to his intuition and to his experience with previous similar designs.

The values of $T_{\text{w,liq}}$ at the throat and at the chamber head, computed by Eq. 6-19 are then compared with the previously determined maximum allowable temperature. If either value is too high or too close to the limit, the remedy might be to reduce q_{liq} or to raise h_{liq} or both. For a given q on the gas side, q_{liq} can be reduced by the use of a finlike structure or even round channels on the cooled side of the chamber wall. The conductance h_{liq} can be raised by making the channel width smaller and increasing the flow velocity, as revealed by Eq. 6-1 with the condition that the volumetric flow rate is kept the same. Because the heat flux is always more intense in the throat region, the duct is usually designed to provide the highest velocity at that section. Thus, the channel width is rarely uniform over the entire motor. There are practical limits to how far h_{liq} can be raised in this way: (1) an increase in flow velocity and a reduction in the channel width results in an increase in pressure loss, which is approximately proportional to the minus 3 to minus 6 power of the width, depending on the type of duct, (2) a reduction in channel width makes the gap tolerances more severe and increases the production cost, and (3) a reduction in channel width makes the width more sensitive to small dimensional changes caused by heating during firing. Of these effects, the pressure loss is usually the most serious, since this imposes a penalty on the propellant feed system. Various design tricks have been used, however, to minimize the pressure loss penalty, but these are too specialized for treatment in this section. For the calculation of pressure loss, the reader is referred to standard works on hydraulics.

Once a satisfactory cooling duct has been designed, the next problem is to decide on the structure of the chamber wall. For the case of a simple, thin shell wall, the gas side wall surface temperature at each section is determined by the given heat flux q and the temperature $T_{w,liq}$ just calculated in the cooling duct design:

$$T_{w,g} = T_{w,liq} + q \frac{d_w}{k_w} \qquad (6\text{-}20)$$

The designer must choose the wall thickness d_w and the wall metal (conductivity k_w). The choice is usually a complicated problem, involving several questions: (1) the strength of the chamber wall and nozzle against the collapsing pressure in the cooling jacket, while at elevated temperature (a short-time creep problem), (2) the chemical resistance of the metal to the hot coolant on one side (corrosion) and to the nonuniform, high velocity hot gas on the other (corrosion and erosion), and (3) the method of fabrication, whether by sheet metal forming and welding or by machining, casting, etc. Several typical combustor designs are discussed in Art. 5. It is obvious, however, that materials of high conductivity and high strength at elevated temperatures are necessary. Alloy steels of the type developed for turbojet blades and disks are usually acceptable, but wall thicknesses in excess of $\frac{1}{8}$ inch are rarely tolerable. In large motors, such thin shells of diameter 1 ft or greater cannot withstand the collapsing pressure, and so devices such as stiffening ribs, support attachments to the outer wall, etc., are employed. Then, these ribs usually do double duty as heat transfer fins and thus reduce the wall temperatures $T_{w,g}$ and $T_{w,liq}$.

This concludes the discussion of the pure regeneratively cooled motor. The following paragraphs present briefly the main ideas of other cooling schemes as well as references to the literature for more complete analyses.

Uncooled heat capacity motor. Except for the injector head which usually incorporates regenerative cooling, the chamber and nozzle are treated as a hollow circular cylinder, using the Fourier equation in cylindrical coordinates at each section, for simplicity, and boundary conditions corresponding to convective heating on the inner surface and total insulation on the outer surface. Solutions of this problem in the following form are obtained:

$$\left(\frac{T_g - T_r}{T_g - T_o} \right) = F \left[\left(\frac{\kappa_w t}{r_o^2} \right), \left(\frac{r}{d_w} \right), \left(\frac{r_o}{r_i} \right), \left(\frac{k_w}{h_g d_w} \right) \right] \qquad (6\text{-}21)$$

The temperature of the metal T_r at the radial distance r, for a cylinder with outside and inside radii r_o and r_i, is given as a function of time t from the start of exposure to the gas of temperature T_g. The metal conductivity k_w, its thermal diffusivity κ_w, and the film conductance in the hot gas h_g enter into the equation. More details of such analyses can be found in V,D.

An approximation frequently employed for the case of thin-walled chambers for such short-duration applications as liquid propellant boosters or anti-aircraft rocket is:

$$\left. \begin{aligned} \rho_w c_w d_w \frac{dT_{av}}{dt} &= h_g(T_g - T_{w_i}) \\ T_{av} &\cong \tfrac{1}{2}(T_{w_i} + T_{w_o}) \\ h_g(T_g - T_{w_i}) &\cong \frac{2k_w(T_{w_i} - T_{w_o})}{d_w} \end{aligned} \right\} \qquad (6\text{-}22)$$

The calculation is then carried out by numerical iteration. The value of h_g can be determined with a sectional, water-cooled motor.

Uncooled motor with refractory insulator. An example of this type of design is the NIKE anti-aircraft guided missile rocket motor. Use of an insulating, high temperature ceramic such as SiC, Al_2O_3, B_4C, or even graphite C usually reduces the heat flux to the outer metal wall to such an extent as to permit a reduction in thickness without loss of strength. The result is an over-all saving of weight. The wall heating problem can be analyzed in a manner similar to that of the simple, unprotected, heat capacity motor, but the final design is usually selected by firing tests. The analytical approach is frustrated by inadequate knowledge of the heat transfer conditions at the rough ceramic surface, of the thermal properties of the ceramic body, of the thermal resistance at the interface between the ceramic and the metal wall, etc.

The selection of a ceramic material is determined largely by considerations other than mere insulating effect: (1) high melting or sublimation temperature of the basic substance, (2) absence of low melting point binders in the ceramic body, (3) chemical resistance to oxidizing or reducing action, depending on the oxygen balance in the combustion gas, (4) low specific gravity, or more generally, low product of specific gravity and thermal conductivity, (5) resistance to thermal shock or "spalling," which in itself requires a high ultimate strength and a low thermal expansivity, and (6) absence of phase transitions with severe density changes within the expected operating range. The usual considerations of availability and cost apply here, particularly since the highest temperature refractories such as B_4C and BeC are available only in small sizes at relatively great expense.

Convective cooling with surface boiling. In the design of the regeneratively cooled motor, the liquid at the heated side of the cooling duct can be prevented from boiling only if the flow velocity is made sufficiently high. As already indicated, this may require an excessive pressure differential or a mechanically unacceptable channel width. It has been found, however, that a slower velocity is allowable, down to a certain lower limit, and in this range boiling occurs on the metal surface (not in the bulk liquid which may have a temperature far below the boiling point). This is the phenomenon of forced convection with surface or "nucleate" boiling discussed in V,E.

When the temperature of the wall adjacent to the liquid is allowed to reach the boiling point at the pressure in the duct, by reducing the velocity, bubbling is seen to begin at the wall. Further reduction of the velocity results in more violent bubbling, but the wall temperature climbs only a few degrees. In effect, the film conductance has been increased by the agitation in the boundary layer. Ultimately, when the velocity has been reduced by 50 per cent or more below the value required to prevent boiling, the vapor formation at the surface has become so rapid that a large part of the surface is blanketed with vapor and the cooling process is disrupted. At this stage, the surface rapidly overheats because of the continuous generation of heat on the hot side of the wall, and an immediate burnout occurs. The burnout heat flux depends on the nature of the liquid, the flow velocity, the degree of subcooling (boiling point minus bulk temperature), the pressure, and the presence of dissolved gases or low boiling point solvents.

The designer usually allows for surface boiling in his design in order to economize on pressure drop, but he is aware of the danger of allowing the velocity to approach too close to the burnout limit.

Cooling with liquefied gases. Rocket motors operating with such propellant combinations as liquid oxygen plus liquid hydrogen, liquid oxygen plus liquid ammonia, liquid fluorine plus liquid diborane, etc., will have to be cooled by a liquefied gas if regenerative cooling is to be employed. In principle, the above discussions concerning convective cooling, with or without surface boiling, apply just as well to this case. Several special problems exist, however, which can be discussed in terms of liquid oxygen as an example.

The atmospheric boiling temperature is 162°R; the critical temperature and pressure are 278°R and 730 lb/in.² abs. If the injection pressure is substantially less than 730 lb/in.² abs, and if the oxygen coolant is admitted to the jacket at 162°R and emerges above the boiling point, the fluid entering the injection orifice will be an uncontrolled mixture of liquid and vapor bubbles, the injection rate will be unsteady, and combustion will be very rough. It is therefore desirable to keep the injection pressure well above 730 psia to avoid bubbling.

There are difficulties in proportioning the dimensions of the cooling duct due to the severe density changes resulting from the progressive warm-up of the supercritical oxygen as it flows along the duct. For example, oxygen at 1000 lb/in.² abs has a density of 71 lb/ft³ at 162°R (at the inlet), 23 lb/ft³ at 300°R, and 7.2 lb/ft³ at 450°R. An additional difficulty revealed by laboratory tests of forced convection with liquid oxygen is that the film conductance goes through a severe minimum in the vicinity of the critical temperature, 278°R. The effect is unexplained.

As yet, there is very little practical experience with liquefied gas cooling of rocket motors.

Internal film cooling. This method has been used in the past mainly as an aid to regenerative cooling, as in the V-2 motor, where it was necessary to incorporate film cooling to rescue a design that was found in the test firings to be inadequately cooled. Liquid, usually the fuel, is injected at low velocity, tangential to the wall if possible, through a ring of injection orifices drilled in the chamber wall just upstream of the overheated zone that needs protection. Ideally, the streams issuing from the orifices collectively form an annular sheet of liquid that flows downstream along the wall. The stream gets progressively thinner as a result of evaporation, until it becomes so thin that surface tension and gas flow disturbances cause break-up of the sheet; this is the end of the protected area, although some protection persists downstream because the boundary layer gas is relatively cool.

It is obvious that the metal wall temperature is no higher than the boiling temperature of the fuel at the pressure in the chamber. As a cooling method, it is very effective. However, there is a specific impulse penalty associated with the imperfect combustion of the fuel that is evaporated into the cool boundary layer, and the fuel stream used to protect the nozzle has no chance to enter the combustion process at all.

A typical magnitude of the flow rate required to protect the inner surface of a nitric acid—JP-4 chamber is 0.005 lb/sec-in.² and about 3 times this rate at the throat. (The fundamental theory to calculate the required flow under various heat transfer conditions, together with the results of experimental researches on film cooling, are described in V,G.) On the basis of this figure, it would take about 5 per cent of the total propellant flow (about 20 per cent of the fuel flow) to fully protect a 10,000-lb-thrust motor in order to dispense entirely with regenerative cooling. The loss in specific impulse would probably be about 3 or 4 per cent.

In general, therefore, film cooling is useful as an auxiliary to, not as a replacement for, regenerative cooling. However, for very large motors, over 100,000-lb thrust, the loss in specific impulse might be less and therefore tolerable, as a result of the lower ratio of surface area to the main propellant flow. Also, for a propellant system such as hydrogen-fluorine with neither substance being suitable as a conventional regenerative coolant, an internal film of hydrogen may be the most practical solution.

Sweat cooling. Some of the practical objections to film cooling using orifice or slot injectors are that perfect surface coverage is difficult to achieve, a large fraction of the injected coolant is ineffective because it fails to evaporate uniformly, and the method is suitable for liquid coolants not for gas coolants. Sweat cooling, that is, admission of the coolant through a porous wall, represents an attempt to get around these objections. In principle, it achieves more uniform coverage and is less wasteful. The theory that permits the calculation of the flow rate required to maintain a desired wall temperature is covered in V,G.

The foremost obstacle to practical use of sweat cooling is the great difficulty of making combustion chambers with the required distribution of permeability, and with the required strength and weight. The original technique was to make the chambers of sintered powdered metal compacts, but this never worked satisfactorily for any but the smallest specimens. A more recent development which has yet to be accepted is spiral winding of fine wires, and then sintering. Another great difficulty is that a porous wall is also an excellent filter for suspended dirt in the fuel; the smallest traces cause local clogging and burnout. Development work in this field is still continuing.

In conclusion, it should be noted that the design of the cooling system for a rocket motor cannot be completely separated from the other aspects of motor design. Thus, a cooling difficulty may sometimes be due to an excessively "hot" injector, and it may be better to try another injector design than to accept extreme difficulties in the cooling system design. Redesign of the chamber shape or exhaust nozzle contour can sometimes alleviate the cooling problem. Also, a small change in the mixture ratio to increase the proportion of the fluid used as the coolant and perhaps to lower the flame temperature may be the best way out of a difficulty. A reduction in combustion pressure may be required if the optimum mixture ratio is to be used. In other words, the design of the motor must be treated as an integrated task.

G,7. Liquid Rocket Systems. The liquid rocket system is the entire complex of devices in the rocket-propelled vehicle that stores, feeds, and burns the liquid propellant. Usually, it is the assignment of the power plant manufacturer to deliver the complete system minus the tanks, instead of merely the rocket motor, because of the necessity for integrated functioning of the entire system. (The power plant manufacturer frequently is concerned with the design of the tanks, too, even if the airframe manufacturer makes them.) Similarly, the designer cannot set the specifications for the rocket motor without regard for the system as a whole.

The components of a liquid rocket system can be classified into four groups according to their functions: the rocket motor assembly, the propellant tanks, the pressurizing equipment, and the engine controls. Many different types of components have been used in liquid rocket systems, to suit the intended purpose. At the end of this article, there is a discussion of the general principles that govern the selection of the optimum system for a particular application.

Compressed gas system. The principal feature that distinguishes a liquid rocket system is the method of pressurizing the propellants for injection into the motor. Although it is a straightforward task to provide the necessary pressure and flow rate in a rocket motor test cell where weight is unimportant and electric power or engine power are unlimited, the problem is much more difficult for a system that must fly. The simplest pressurizing scheme is the regulated compressed gas type. Having no moving machinery other than valves, it has the advantage of great reliability. Its disadvantages can be serious, however, for certain applications. First, its weight increases rapidly with firing duration in a linear manner, due to the proportional increase in the volume and weight of the propellant tanks and the gas tank. This weight is prohibitive for durations longer than about 30 to 60 seconds. Second, the tanks must be spherical or cylindrical to withstand the high pressures, which sometimes complicates their placement in an airplane or other vehicle, although cylindrical tanks can be accommodated nicely in guided missiles or high altitude research rockets. Third, under some circumstances, the field problem of supplying high pressure gas is troublesome.

The principal components are the gas pressure tank, the propellant tanks, the pressure regulator(s), the gas starting valve, and the propellant flow control valves.

The pressure regulator is a servo-operated metering valve that senses the delivery pressure, compares it with the predetermined pressure setting, and so throttles the outflow of gas as to keep the difference nearly zero. The design of high volume regulators is a specialized task, but several of its performance characteristics deserve mention. A regulator, like any servomechanism with feedback, can hunt or "chatter"; introduction of damping or modification of the orifice and pintle shapes may be required. Condensation or freezing of moisture or oil, if present in the gas supply, or even of the gas itself, may occur on the seat as a result of expansive cooling. The delivery pressure may climb or fall appreciably, as the inlet pressure diminishes during the course of a firing, because of inexact balance of pressures on the throttling pintle. Finally, if the propellant valves are shut off in the middle of a firing, the regulated pressure may climb considerably above that delivered during normal operation, due to the large actuating force required to close the valve tightly enough to stop the flow.

The gas starting valve located upstream of the regulator provides a secure means of retaining the high pressure gas until the start of firing, and serves to shut off the gas flow when necessary to avoid any dangerously high "lock-up" pressure delivered by the regulator.

The propellant flow control valves are of various types and serve several functions. A positive seal before the start of firing is provided by frangible diaphragm units, designed to burst open at some fraction of the starting feed pressure. Throttling valves are provided to gradually increase the flow rates according to a definite program during the start-up transient. Shut-off valves are provided in one or both lines to terminate the thrust promptly if desired. Many additional valves appear in even the simplest rocket system that have to do with filling the propellant tanks, charging the gas tank, venting the tanks, draining the system at various points, checking reverse flow, relieving excessive pressure, etc.

Several considerations determine the volume of the compressed gas tank. First is the charging pressure: the higher the pressure, the more compact the tank, but at the same time the more difficult the logistic supply problem. Pressures between 2000 and 5000 lb/in.^{2} have been used, with the current trend toward the higher pressures as suitable sources become available. Second is the type of gas: it must be available at the required pressure at the launching site; it must not react with either propellant or dissolve too easily in them; it must not condense in the throttle of the regulator; it should have the lowest possible density. These considerations rule out, for example, hydrogen or carbon dioxide, but favor helium, nitrogen, or air.

The ideal throttling process is isothermal, for a perfect gas like helium, and is nearly so except for a small Joule-Thompson drop for nitrogen and air. The expansive cooling mentioned above is localized in the high velocity gas region of the throttle, where the process is nearly isentropic, but the temperature at the outlet is nearly equal to the inlet temperature. The inlet temperature falls steadily from the start of firing, if the system is adiabatic, as a consequence of the isentropic expansion in the gas tank. On the basis of this description, it is possible to integrate the gas flow rate to determine the mass of gas m_g needed in a gas tank of volume $\mathcal{U}_g$ to displace the propellants and fill the propellant tanks of volume $\mathcal{U}_p$ with gas at regulated pressure p_p, when the gas tank is charged at p_1 and runs down to pressure p_2:

$$m_g = \frac{p_p \mathcal{U}_p \mathfrak{M}}{RT_1} \left(\frac{\gamma}{1 - p_2/p_1} \right) = \frac{p_1 \mathcal{U}_g \mathfrak{M}}{RT_1} \tag{7-1}$$

If the over-all expansion were isothermal in the tank as well as in the regulator, and if all the gas in the gas tank were used in the propellant tanks, the quantity in brackets would be unity, and the remaining factor is easily recognized as the result of Boyle's law. In practice, the actual m_g can be somewhat less than Eq. 7-1 since some heat is actually transferred to the gas during the expansion, and since it is quite feasible to permit p_2 and p_p to fall below the nominal design levels in the last part of the firing period. The above formula makes no allowance, of course, for gas that dissolves in the propellants.

A typical figure for the mass of air required for a rocket like the Aerobee is 5 per cent of the propellant mass; for helium, it would be about 1 per cent. The switch from air to helium is obviously desirable for high performance missiles and research rockets, and especially so because of the solubility problem in the case of liquid oxygen systems. Additional savings in the inert gas weight can be effected by warming the gas before it is admitted to the propellant tank, by allowing it to absorb some of the heat of the rocket motor in a suitable heat exchanger. An important saving in the weight of the gas tank is possible in principle, but complicated in practice, by charging with liquefied gas into a well-insulated gas tank and gasifying it with an auxiliary heater when the gas is to flow into the propellant tanks.

Turbopump system. The weight of the gas tank, the pressure-stressed propellant tanks, and the gas in the compressed gas system usually amounts to about 15 to 20 per cent of the propellant weight. For some applications this weight penalty can be reduced sharply by using a pumping system with lightweight, thin shell, propellant tanks. The most common power source to drive the propellant pumps is a chemical gas-driven turbine, the combination comprising a turbopump. The following considerations determine the design of a turbopump.

The turbine shaft power required to drive the pumps to supply the rocket motors can be calculated as follows:

$$P_t = P_f + P_o = \frac{\Delta p_f \dot{m}_f / \rho_f}{\eta_f} + \frac{\Delta p_o \dot{m}_o / \rho_o}{\eta_o} \qquad (7\text{-}2)$$

The pump efficiencies η_f and η_o are defined as the ratio of hydraulic output power to shaft input power for the fuel and oxidizer flows respectively, as implied in Eq. 7-2. The efficiency of a pump can be determined in advance in a water test flow system; it varies over a wide range according to the type of design and the particular operating point. For rocket pumps, with their particular limitations, an efficiency of 70 per cent is quite good, and 40 per cent may be acceptable.

The driving turbine is usually a simple-, single-, or two-stage, impulse-type unit, driven by gases generated at a moderate temperature by decomposition or combustion of a propellant. The mass flow $\dot{m}_t$ consumed by the turbine to deliver a shaft power P_t can be calculated according to a measured or computed efficiency η_t:

$$P_t = \dot{m}_t(h_1 - h_2)_a = \eta_t \dot{m}_t(h_1 - h_2)_i \cong \eta_t \dot{m}_t \overline{C}_p T_1 \left[1 - \left(\frac{p_2}{p_1}\right)^{\frac{\gamma-1}{\gamma}} \right] \qquad (7\text{-}3)$$

In this equation, the ideal enthalpy drop $(h_1 - h_2)_i$ is replaced by the equivalent expression for a near-perfect gas that enters at pressure p_1

and temperature T_1 and leaves at pressure p_2 after an isentropic work process. The departure from isentropicity is measured by the efficiency η_t, which indicates the fraction of the ideal enthalpy drop that actually occurs in the turbine, $(h_1 - h_2)_s$. To carry out precise turbine performance calculations, a Mollier diagram is required. Typical values of η_t in existing rocket engines range from 30 to 40 per cent. These relatively low values are the result of adopting simple one- or two-stage turbines to minimize the fixed installation weight; the penalty is a higher rate of turbine propellant consumption and a larger gas generator. Some designers have kept the weight down also by using aluminum alloy turbines with aluminum blades, even though this necessitated a limit on the temperature of the driving gas of about 725°F.

From Eq. 7-2 and 7-3, it is clear that the mass flow of turbine propellant varies inversely as the products $\eta_p\eta_t$ and $\eta_o\eta_t$ and directly as the feed pressure. Typical values of turbine flow are as low as 1.5 per cent of the rocket engine flow for large missiles, and as much as 5 per cent for engines of less than 10,000-lb thrust. A large part of the difference results from the wide variations in pump and turbine efficiencies, which depend on optimization studies that seek to balance the larger dead weight of an efficient turbopump against the higher consumption rate of an inefficient one. Typical turbine powers are about 0.01 to 0.02 hp/lb thrust. The V-2 turbopumps delivered 465 hp at rated operating conditions.

Inasmuch as the over-all specific impulse of a rocket engine must be based on the total propellant flow which includes the turbine flow, it is clear that turbopump performance analysis is just as important as the analysis given in Art. 3 of the corrections to ideal rocket motor theory. For this purpose, the reader is referred to X,F for the principles of axial flow turbines and to X,J for the principles of centrifugal compressors. Certain special design features of rocket turbopumps deserve mention here, however.

Light weight is achieved by the use of a single shaft, without speed changer, to drive the pump impellers. The high rotary speed demanded by the single-stage turbine usually means that a single centrifugal impeller is adequate to develop the full pressure required by the rocket engine. Unfortunately, the development of high pressure in a single high speed impeller of small diameter means a high relative inlet velocity at the impeller eye and a strong tendency toward cavitation. Cavitation, the boiling of the liquid on the pump vanes and in the inlet passage due to the reduced static pressure, results in a reduction in fluid delivery and, if it is extensive, in damaging vibration. To suppress cavitation, it is necessary to boost the pressure at the pump inlet. Booster pumps, if they are not to cavitate themselves, must be slow speed, large, and therefore heavy devices, and need a driving power source. The alternative frequently employed is gas pressurization of the tanks to about 5 or 10 per cent of the main pump pressure. The choice between booster pumps and gas pressurization depends largely on the necessary tank configuration and is decided by a weight optimization analysis.

The hot gas supply to drive the turbine is created by a combustion or decomposition reaction in a gas generator or "combustion pot." The most commonly used gas generation propellant is concentrated hydrogen peroxide. The range of turbine inlet temperatures that can be obtained with various water concentrations in the peroxide is as follows: For aluminum alloy turbines, the maximum allowable temperature is about 1000°F; for steel alloys, about 2000°F. (These are somewhat higher than the temperatures allowed for relatively long-life turbomachines.) The gas generator system is quite similar to a rocket system, involving a propellant tank, a pressurizing device, flow controls, as well as a combustion chamber. Propellants other than concentrated hydrogen peroxide have been used or proposed as gas generants.

The hot gas that leaves the driving turbine can either be discharged rearward with moderate back pressure to add a small amount of thrust to the engine, or it can be used as a heat source to vaporize, for example, a liquefied pressurizing gas, or it can be used to actuate some devices in the flight control or engine control system. The back pressure can usually be accommodated with a small loss of turbine efficiency.

Auxiliary devices are sometimes connected to the driving shaft of the turbine, for example, an electric generator for large electric power requirements or a hydraulic pump to actuate various controls. The extra propellant consumption to provide this power must be included in the over-all specific impulse computation.

The first successful U. S. turbopump-type rocket engine, developed in 1945, operated on nitric acid with an aniline-furfuryl alcohol mixture as the fuel. The driving gas was generated by burning the same propellants together with a large diluent flow of water. The German ME-163 rocket interceptor plane was propelled by a turbopump-type engine in which the driving gas was produced by hydrogen peroxide decomposition and the propellants were peroxide and a hydrazine-alcohol-water mixture.

Other systems. It was pointed out above that the main distinguishing feature of a liquid rocket system is the propellant pressurizing scheme. In addition to the two principal methods based on pressurized gas and turbopumps, other methods have been proposed or actually used in one or more rockets:

1. A solid propellant cartridge can be used instead of the high pressure gas source, the propellants being pressurized by the combustion products. A system of this type was employed in the German anti-aircraft "Taifun."
2. Controlled combustion can be allowed to take place in the fuel tank by

injecting a small spray of oxidizer, and vice versa, thereby creating pressure in the tanks. This has been tried experimentally, but never used in practice. Similarly, the gases produced in a separate gas generator can be used directly to pressurize the propellants.

3. A novel version of the turbopump system is the so-called jet blast type, in which the turbine is driven by the rocket exhaust jet. In the only fully developed unit of this type, the turbine is a single, high speed, axial flow wheel, only partially immersed in the hot blast. This partial immersion allows air cooling of the blades. Although it has the advantage of very light weight, a number of serious starting, regulating, and operating difficulties have put it in disfavor.

4. Another ingenious mechanism is the so-called rotojet with the rocket motors mounted on a rotatable shaft at a small angle, so as to create rotary shaft power to drive the pumps when thrust is being delivered. The development was not successful, due to mechanical difficulties, sealing problems, and erratic combustion. A similar device, running at even higher speed to create the required feed pressure by centrifugal action, was developed experimentally but ran into comparable difficulties.

5. It has been proposed to use the injector principle commonly used to inject feed-water into steam boilers, the propellants to be heated in a boiler surrounding the rocket motor. The starting problem and the difficulties of heating propellants to high temperature seem to make this scheme impractical.

6. Various types of mechanical power can be employed to drive the propellant pumps, depending on the conditions of the application. An auxiliary rocket engine for a turbojet airplane can draw pumping power from the main gas turbines by bleeding air from the compressor discharge to run an air turbine. Similarly, the pumps can be mounted as accessories on the gear box of the engine, if it is of the reciprocating type. A JATO rocket engine of this type was used in World War II. Finally, a separate auxiliary gas turbine or reciprocating engine can be used to power the pumps.

System selection. The optimum liquid rocket system for a particular application is determined on the basis of the following considerations:

1. Minimum weight: In the competition for minimum weight between the two main system types, compressed gas and turbopump, it is evident that the firing duration is the principal controlling factor. For short durations, the large fixed weight of the turbopump and gas generation system is a penalty; for long durations, the very heavy tank system of the compressed gas type is objectionable. The cross-over point is about 15 seconds for high thrust engines, 30 seconds for small engines. This comparison can be made only after each system is optimized

with respect to its operating parameters. The most important of these is the combustion pressure—the higher the pressure, the lower the propellant consumption rate for a given thrust; but at the same time, the greater the weight of the pressurization system. For the compressed gas type, the optimum pressure usually works out to lie between 250 and 400 lb/in.²; for the turbopump system, the optimum is usually near 1000 lb/in.², but a lower pressure is always employed to ease the cooling problem. It is clear that the turbopump type is the most suitable one for long range, large size rockets with long firing durations.

2. Simplicity and ruggedness: The great simplicity of the compressed gas type makes it the most appropriate system for such applications as small anti-aircraft guided missiles, which have to be ready for instantaneous firing with high reliability with a minimum of prefiring inspection and adjustment. The same consideration applies to detachable JATO units.

3. Low cost: The consideration of cost is important mainly for civilian applications such as the take-off of heavy cargo planes, and perhaps rocket travel in the future. For military purposes, cost is significant mainly as an indication of manufacturing difficulty, and, therefore, for missiles or JATOs required in very large numbers, the gas-pressurized type is preferred.

G,8. Bibliography.

Article 1.

Dictionary of guided missile terms. Compiled by Ordnance Dept., U.S. Army. *Anti-Aircraft J.*, 1949.
Glossary of guided missile terms. Prepared by *Committee on Guided Missiles, Research and Develop. Board, U.S. Dept. of Defense, GM 51/3 and GM 51/8*, Sept. 1948.
Layton, J. P., and Youngquist, R. Proposed A.S.A. letter symbols for rocket propulsion. *Jet Propulsion 25*, 634 (1955).

Article 2.

Malina, F. J. Characteristics of the rocket motor unit based on the theory of perfect gases. *J. Franklin Inst. 230*, 433 (1940).
Seifert, H. S., and Crum, J. Thrust coefficient and expansion ratio tables. *Ramo-Wooldridge Corp.*, Feb. 1956.
Seifert, H. S., Mills, M. M., and Summerfield, M. The physics of rockets, Part I. *Am. J. Phys. 15*, 1 (1947).
Shapiro, A. H. *Dynamics and Thermodynamics of Compressible Fluid Flow*, Vol. I, Chap. 4. Ronald Press, 1953.
Sutton, G. P. *Rocket Propulsion Elements*, 2nd ed., Chap. 3. Wiley, 1956.

Article 3.

Banerian, G. Rocket performance with heat added to inlet propellants by regenerative and external means. *Jet Propul. 25*, 712 (1955).
Beckwith, I. E., and Moore, J. A. An accurate and rapid method for the design of supersonic nozzles. *NACA Tech. Note 3322*, Feb. 1955.

Dillon, P., and Line, L. E., Jr. Heat transfer between solid particles and gas in a rocket nozzle. *Jet Propul. 26*, 1091 (1956).

Durham, F. P. Thrust characteristics of underexpanded nozzles. *Jet Propul. 25*, 696 (1955).

Gilbert, M., Davis, L., and Altman, D. Velocity lag of particles in linearly accelerated combustion gases. *J. Am. Rocket Soc. 25*, 26–30 (1955).

Glassman, I. Impulse expressions for rocket systems containing a solid phase. *Jet Propul. 27*, 542 (1957).

Kogan, A. Boundary layer correction in supersonic nozzle scaling. *J. Aeronaut. Sci. 25*, 64 (1958).

Puckett, A. E. Supersonic nozzle design. *J. Applied Mech. 13*, A265-A270 (1946).

Seifert, H. S., and Altman, D. A comparison of adiabatic and isothermal expansion processes in rocket nozzles. *J. Am. Rocket Soc. 22*, 159–162 (1952).

Shapiro, A. H. *The Dynamics and Thermodynamics of Compressible Fluid Flow*, Vol. I, Chap. 7 and 15. Ronald Press, 1953.

Summerfield, M., Foster, C. R., and Swan, W. C. Flow separation in overexpanded supersonic exhaust nozzles. *Jet Propul. 24*, 319 (1954).

Sutton, G. P. *Rocket Propulsion Elements*, 2nd ed., Chap. 3. Wiley, 1956.

Article 4.

Battelle Memorial Inst. Physical properties and thermodynamic functions of fuels, oxidizers, and products of combustion. *Project Rand Repts. R-127, 129, 196.* Columbus, Ohio, 1949.

Charts of theoretical performance of several rocket propellant combinations. *Rocketdyne, Div. North Amer. Aviation, Inc.*, 1956.

Donegan, A. J., and Farber, M. Solution of thermochemical propellant calculations on a high speed digital computer. *Jet Propul. 26*, 164–172 (1956).

Fickett, W., and Cowan, R. D. Values of thermodynamic functions to 12,000°K for several substances. *Los Alamos Sci. Lab. Rept. 1727*, Sept. 1954.

Hottel, H. C., Williams, G. C., and Satterfield, C. N. *Thermodynamic Charts for Combustion Processes.* Wiley, 1949.

Huff, V. N., and Calvert, C. S. Charts for computation of equilibrium composition of chemical reactions in the C-H-O-N system from 2000°K to 5000°K. *NACA Tech. Note 1653*, July 1948.

Huff, V. N., and Gordon, S. Tables of thermodynamic functions for analysis of aircraft propulsion systems. *NACA Tech. Note 2161*, Aug. 1950. (Includes light metals and halogens, as well as C, H, O, and N compounds.)

Huff, V. N., and Morrell, V. E. General method for computation of equilibrium composition and temperature of chemical reactions. *NACA Tech. Note 2113*, June 1950.

Krieger, F. J. Chemical kinetics and rocket nozzle design. *J. Am. Rocket Soc. 21*, 179–185 (1951).

Lewis, B., and von Elbe, G. *Combustion, Flames and Explosions of Gases*, Chap. 13. Academic Press, 1951.

Natl. Bur. Standards. Selected values of properties of hydrocarbons. *Circular 461*, Gov't. Printing Office, 1947.

Natl. Bur. Standards. Tables of selected values of chemical thermodynamic properties, 3 vol. *Circular 500*, Gov't. Printing Office, 1952.

Penner, S. S. Thermodynamics and chemical kinetics of one-dimensional nonviscous flow through a Laval nozzle. *J. Chem. Phys. 19*, July 1951.

Pocket Data Book for Rocket Engines. Bell Aircraft Corp., Buffalo, N.Y., 1954. (Properties of fuels, oxidizers, and propellant combinations.)

Sutton, G. P., *Rocket Propulsion Elements*, 2nd ed., Chap. 4. Wiley, 1956.

Vichnievsky, R., Sale, B., and Marcadet, J. Combustion temperatures and gas composition. *J. Am. Rocket Soc. 25*, 105 (1955).

Article 5.

Baker, D. I. Mixture ratio and temperature survey of ammonia-oxygen rocket motor combustion chambers. *Jet Propul. 25*, 217 (1955).

Barrere, M., and Moutet, A. Instabilities of low frequency in a rocket motor. *Recherche aéronaut. Paris 44*, 29 (1955).

Baxter, A. D. Combustion in the rocket motor. *J. Brit. Interplanetary Soc. 10*, May 1951.

Berman, K., and Cheney, S. H. *Rocket Motor Instability Studies 25*, 513 (1955).

Crocco, L. Aspects of combustion instability in liquid propellant rocket motors. *J. Am. Rocket Soc. 21*, 163 (1951); *22*, 7 (1952).

Crocco, L. Considerations on the problem of scaling rocket motors. *Selected Combustion Problems*, Vol. 2. Butterworths, London, 1956.

Crocco, L., and Cheng, S. I. *Theory of Combustion Instability in Liquid Propellant Rocket Motors*. Butterworths, London, 1956.

Diplock, B. R., Lofts, D. L., and Grimston, R. A. Theory, design, and development of liquid propellant rocket motors. *J. Roy. Aeronaut. Soc. 57*, Jan. 1953.

Gunn, S. V. The effects of several variables upon the ignition lag of hypergolic fuels oxidized by nitric acid. *J. Am. Rocket Soc. 22*, 33–38 (1952).

Penner, S. S., and Fuhs, A. E. On generalized scaling procedures for liquid fuel rocket engines. *Combustion and Flame Quart. 1*, June 1957.

Ross, C. C. Scaling of liquid fuel rocket combustion chambers. *Selected Combustion Problems*, Vol. 2. Butterworths, London, 1956.

Ross, C. C., and Datner, P. P. Combustion instability in liquid propellant rocket motors—A survey. *Selected Combustion Problems*, Vol. 1. Butterworths, London, 1954.

Stehling, K. R. Injector spray and hydraulic factors in rocket motor analysis. *J. Am. Rocket Soc. 22*, 132–138 (1952).

Summerfield, M. A theory of unstable combustion in liquid propellant rocket motors. *J. Am. Rocket Soc. 21*, 108 (1951).

Sutton, G. P. *Rocket Propulsion Elements*, 2nd ed., Chap. 7. Wiley, 1956.

Tischler, A. O., and Bellman, D. R. Combustion instability in an acid-heptane rocket with a pressurized gas propellant pumping system. *NACA Tech. Note 2936*, 1953.

Tischler, A. O., and Dalgleish, J. E. Experimental investigation of a lightweight rocket chamber. *NACA Research Mem. E52L19A*, Dec. 1952.

Trent, C. H. Investigation of combustion in rocket combustion chambers. *Ind. Eng. Chem. 48*, 749 (1956).

Tsien, H. S. Servo-stabilization of combustion in rocket motors. *J. Am. Rocket Soc. 22*, 256 (1952).

Article 6.

Bartz, D. R. A simple equation for rapid estimation of rocket nozzle convective heat transfer coefficients. *Jet Propul. 27*, 49 (1957).

Boden, R. H. Heat transfer in rocket motors and the application of film and sweat cooling. *Trans. Am. Soc. Mech. Engrs. 73*, May 1951.

Cheng, C. M. Resistance to thermal shock. *J. Am. Rocket Soc. 21*, 147 (1951).

Dunn, L. G., Powell, W. B., and Seifert, H. S. Heat transfer studies relating to rocket power plant development. *Proc. Third Anglo-American Aeronaut. Conference*, Brighton, England, 1951. (Published by Roy. Aeronaut. Soc.)

Duwez, P. and Wheeler, H. L., Jr. Experimental study of cooling by injection of a fluid through a porous material. *J. Aeronaut. Sci. 15*, 509–521 (1948).

Eckert, E. R. G. Convective heat transfer at high velocities. *Heat Transfer Symposium*, Univ. Mich. Press, 1952.

Eckert, E. R. G. Transpiration and film cooling. *Heat Transfer Symposium*, Univ. Mich. Press, 1952.

Ellion, M. E. New technique for obtaining heat-transfer parameters of the wall and combustion gas in a rocket motor. *Trans. Am. Soc. Mech. Engrs. 73*, 109–114 (1951).

Elsasser, W. M. Mean absorption and equivalent absorption coefficient of a band spectrum. *Phys. Rev. 54*, 126 (1938).

Geckler, R. D. Transient, radial heat conduction in hollow circular cylinders. *Jet Propul. 25*, 31–35 (1955).

Greenfield, S. Determination of rocket-motor heat-transfer coefficients by the transient method. *J. Aeronaut. Sci. 18*, 512–518 (1951).

Jakob, M. *Heat Transfer*, Vol. I, Chap. 4. Wiley, 1949.

John, R. R., and Summerfield, M. Effect of turbulence on radiation intensity from propane-air flames. *Jet Propul. 27*, 169 (1957).

Kennard, E. H. *Kinetic Theory of Gases.* McGraw-Hill, 1938.

Kreith, F. and Summerfield, M. Heat transfer to water at high flux densities with and without surface boiling. *Trans. Am. Soc. Mech. Enjrs., 71*, 805 (1949).

Manson, S. S. Behavior of materials under conditions of thermal stress. *Heat Transfer Symposium*, Univ. Mich. Press, 1952.

McAdams, W. H., *Heat Transmission*, 3rd ed., Chap. 4, 7, and 8. McGraw-Hill, 1954.

Natl. Bur. Standards. Tables of thermal properties of gases. *Circular 564*, U.S. Gov't. Printing Office, 1955.

Penner, S. S. Emissivity calculations for diatomic gases. *J. Appl. Mech. 18*, 53 (1951).

Rohsenow, W. M. Heat transfer with evaporation. *Heat Transfer Symposium*, Univ. Mich. Press, 1952.

Saunders, O. A. Heat transfer on a nozzle at supersonic speeds. *Engineering 173*, Aug. 1952.

Summerfield, M. Fundamental problems in rocket research. *J. Am. Rocket Soc. 81*, 79–98 (1950).

Summerfield, M. Recent developments in convective heat transfer with special reference to high temperature combustion chambers. *Heat Transfer Symposium*, Univ. Mich. Press, 1952.

Sutton, G. P. *Rocket Propulsion Elements*, 2nd ed., Chap. 7. Wiley, 1956.

Symposium on Transport Properties in Gases. Northwestern Univ. Press, 1958.

Von Bahr, E. Influence of pressure on the absorption of infra-red rays by gases. *Ann. Physik 33*, 585 (1910).

von Karman, Th. The analogy between fluid friction and heat transfer. *Trans. Am. Soc. Mech. Engrs. 61*, 705 (1939).

Article 7

Gore, M. R., and Carroll, J. J. Dynamics of a variable thrust, pump-fed, bi-propellant liquid rocket engine system. *Jet Propul. 27*, 35 (1957).

Reichel, R. H. The importance of mixture ratio control for large rocket vehicles. *Jet Propul. 25*, 291–293 (1955).

Ross, C. C. Principles of rocket-turbopump design. *J. Am. Rocket Soc. 84*, 21–33 (1951).

Stehling, K. R., and Diamond, P. M. Flow controls. *J. Am. Rocket Soc. 23*, 178–183 (1953).

Sutton, G. P. *Rocket Propulsion Elements*, 2nd ed., Chap. 8. Wiley, 1956.

Tsien, H. S. *Engineering Cybernetics*, Chap. 5 and 8. McGraw-Hill, 1954.

Milton Keynes UK
Ingram Content Group UK Ltd.
UKHW021818250823
427506UK00006B/161